International Conference on Virtual and Mixed Reality Interfaces 2020

Held online due to COVID-19

IOP Conference Series: Materials Science and Engineering
Volume 979

Johor, Malaysia
16 - 17 November 2020

ISBN: 978-1-7138-3179-2
ISSN: 1757-8981

Printed from e-media with permission by:

Curran Associates, Inc.
57 Morehouse Lane
Red Hook, NY 12571

Some format issues inherent in the e-media version may also appear in this print version.

Institute of Physics
Dirac House, Temple Back
Bristol BS1 6BE UK

Phone: 44 1 17 929 7481
Fax: 44 1 17 920 0979

techtracking@iop.org

Additional copies of this publication are available from:

Curran Associates, Inc.
57 Morehouse Lane
Red Hook, NY 12571 USA
Phone: 845-758-0400
Fax: 845-758-2633
Email: curran@proceedings.com
Web: www.proceedings.com

TABLE OF CONTENTS

IOP Conf. Series: Materials Science and Engineering **979** (2020) 011001 doi:10.1088/1757-899X/979/1/011001

Preface

The rise of new technology for Industry 4.0, we aim to discover how Virtual and Mixed Reality Interfaces fit to Industry 4.0. International Conference on Virtual and Mixed Reality Interfaces 2020 (ICVRMR2020) brings an opportunity to invite new innovative inventions on Computer Graphics, Multimedia, Image Processing, Human Computer Interaction, Animation, Web and Mobile Technology, Games and Emerging Technology especially in Virtual and Augmented/Mixed Reality. Due to the COVID-19 pandemic situation all around the globe, we have organized this conference into virtual conference and the method is virtual presentation for all International and local participants. ICVRMR2020 is a strategic step to develop a research network in the field of Computer Graphics, Multimedia, Image Processing, Human Computer Interaction, Animation, Web and Mobile Technology, Games and Emerging Technology especially in Virtual and Augmented/Mixed Reality. UTM ViCubelab Research Group brings this virtual conference to provide the opportunities for academicians, professionals, practitioners and policy makers in the engineering and computing fields to share their thoughts and empirical works on Virtual Reality and Augmented/Mixed Reality inline with the rise of new industrial digital technology. Therefore, we still held this conference even though it is virtually through the Webex application. Outcome from this conference can be a benchmark for future research and development, and become a trigger point for developing research and development collaborations. It also serves to foster communication among researchers and practitioners working in a wide variety of scientific areas with a common interest in Engineering and Applied Computing.

Thank you.

Dr. Ajune Wanis Ismail

Chair of ICVRMR 2020

IOP Conf. Series: Materials Science and Engineering **979** (2020) 011001 doi:10.1088/1757-899X/979/1/011001

Committee Members

Honorary Chair

Prof. Dato' Ir. Dr. Mohammed Rafiq Bin Abdul Kadir (Dean, Faculty of Engineering UTM)

Chair

Dr Ajune Wanis Ismail (Universiti Teknologi Malaysia)

Co-chair

Professor Dr Mark Billinghurst (University of South Australia)

Publication Chair

Dr Norhaida Mohd Suaib (Universiti Teknologi Malaysia)

Our Committee

Prof Dr Mohd Shafry Mohd Rahim (Universiti Teknologi Malaysia)

Assoc. Prof. Dr. Abdullah Bade (Universiti Malaysia Sabah)

Assoc. Prof. Dr. Ing. Maizatul Hayati Mohamad Yatim (Universiti Pendidikan Sultan Idris)

Assoc. Prof. Dr. Nor Azman Ismail (Universiti Teknologi Malaysia)

Assoc. Prof. Dr. Masitah Ghazali (Universiti Teknologi Malaysia)

Dr. Adrian Clark (University of Canterbury)

Dr. Farhan Mohamed (Universiti Teknologi Malaysia)

Dr. Hamzah Asyrani Sulaiman (Universiti Teknikal Malaysia Melaka)

Dr. Ibrahim Ahmad (Universiti Teknikal Malaysia Melaka)

Dr. Jumail Taliba (Universiti Teknologi Malaysia)

Dr. Ismahafezi Ismail (Universiti Sultan Zainal Abidin)

Dr. Khurram Ejaz (University of Gujrat)

Dr. Md Sah Hj Salam (Universiti Teknologi Malaysia)

Dr. Mohamad Faizal Ab Jabal (Universiti Teknologi MARA)

Dr. Noor Dayana Binti Abd Halim (Universiti Teknologi Malaysia)

Dr. Nur Saadah Mohd Shapri (Universiti Sultan Zainal Abidin)

Dr. Suriati Sadimon (Universiti Teknologi Malaysia)

Dr. Tan Chi Wee (Tunku Abdul Rahman University College)

Dr. Yusliza Yusoff (Universiti Teknologi Malaysia)

Dr. Yacine Izza (Toulouse University)

Cik Suhaimi Yusof (Universiti Teknologi Malaysia)

Nor Anita Fairos Ismail (Universiti Teknikal Malaysia Melaka)

Iznora Aini Zolkipli (UNITAR International University)

Mohamad Yahya Fekri (Universiti Teknologi Malaysia)

Muhammad Nur Affendy (Universiti Teknologi Malaysia)

Nazreen Abdullasim (Universiti Teknikal Malaysia Melaka)

Rosely Kumoi (Universiti Teknologi Malaysia)

Ahmad Fariz Ali (Universiti Teknologi Malaysia)

IOP Conf. Series: Materials Science and Engineering **979** (2020) 011002 doi:10.1088/1757-899X/979/1/011002

Peer review declaration

All papers published in this volume of *IOP Conference Series: Materials Science and Engineering* have been peer reviewed through processes administered by the Editors. Reviews were conducted by expert referees to the professional and scientific standards expected of a proceedings journal published by IOP Publishing.

- **Type of peer review:** ***Single-blind***
- **Conference submission management system:** ***Easychair.org***
- **Number of submissions received:** ***42***
- **Number of submissions sent for review:** ***40***
- **Number of submissions accepted:** ***19***
- **Acceptance Rate (Number of Submissions Accepted / Number of Submissions Received X 100):** ***47.5%***
- **Average number of reviews per paper:** ***2 to 3 reviews***
- **Total number of reviewers involved:** ***22***
- **Any additional info on review process:** ***Preliminary Review - The papers that passed the first and second reviews will be reviewed again to suit the following aspects: Originality and Clarity***
- **Contact person for queries:** ***Ajune Wanis Ismail, ajune@utm.my***

IOP Conf. Series: Materials Science and Engineering **979** (2020) 012001 doi:10.1088/1757-899X/979/1/012001

ARGarden: 3D outdoor landscape design using handheld augmented reality with multi-user interaction

F E Fadzli [1,2] *, M A Mohd Yusof [1,2], A W Ismail [1,2], M S Hj Salam [1,2] and N A Ismail [1,2]

[1] Mixed and Virtual Environment Research Lab (mivielab), ViCubeLab, Universiti Teknologi Malaysia, 81310 Johore, Malaysia
[2] School of Computing, Faculty of Engineering, Universiti Teknologi Malaysia, 81310 Johor, Malaysia

*fazliaty.edora@gmail.com

Abstract. Augmented Reality (AR) is a technology to display the virtual object by overlaying on the physical marker in the real environment. Nowadays, the handheld device becomes a trend in AR since it has high processing power and compatible with the AR system. However, most of the AR applications did not fully support the interaction from multiple users for the collaborative AR interface. Therefore, this research aim is to develop multi-user interaction for collaborative handheld AR. There are a few phases involved to design and develop the collaborative AR application for the handheld device called ARGarden. All the interaction that has been done for the virtual object transformation such as translation, rotation and scaling can be seen by the other users, and this data is synchronized by using a network protocol. Then, a queue system is implemented for the multi-user interaction to allow the virtual object can only be selected by one user at one time, and the other users must wait for their turn. The significance of this research is all the method that has been implemented for the ARGarden can be applied in different fields such as simulation, modelling, medical and education.

1. Introduction

Virtual reality (VR) or virtual environment is an intuitive way to enable the participant to accomplish the task in the virtual world which impossible in the real world by creating the synthesis of reality [1]. This technology makes the user feel they self-inside the VR, and this is called immersive. Immersive happen when there is realism, low latency which is, the time taken for refresh is around 17 milliseconds and smooth, which the refresh rate is about 80 to 120 Hz and framerate per second is 60 fps for GPU and 30 fps for CPU. However, the purpose of augmented reality (AR) is the opposite of VR. VR is used to create the virtual world and replace reality, but AR is used to enhance reality. In order to create mixed reality (MR) in real-time, AR technology refers to placing the virtual elements in the actual real environments view to improve and enhance the senses of human perceptions in the real world [2]. Therefore, AR can improve user interaction with the real world by displaying important information that cannot be recognized or identified directly by using their senses. This important information can be in virtual, for example, mobile augmented reality shopping application for Dulux Paint Expert which enables the user to see the realistic Dulux Trade paint colours appear on the wall indirectly give them the paint information [3].

Nowadays, many prefer to use AR because it helps them to understand some information much better rather than look at the direct view of the information. For example, the usage of VR and AR in dentistry not only uses the system for learning and teaching from their perspective but also training their skill and improve the hand-eye coordinate indirectly help the user to correct the posture and skill [4]. There is also an AR application for modelling 3D object in real environment [5].

In AR, there are a few problems that are still unsolved until today. Most of the problem usually related to the fundamental of AR, which consists of four categories, display technique, tracking technique, interface, and interaction. To enable the user to seamlessly interact with the virtual object by implementing the most suitable interaction technique for AR application is one of the essential research in AR [6]. AR interface often only for the view, and the users cannot manipulate. Then, the interaction comes. Instead of only can browse the AR by the user, with the interaction the user can manipulate the virtual element in the AR. The interaction must be intuitive and be natural as possible to enhance the natural user interface (NUI). Besides that, there are four techniques to display AR, such as a projector, depth-HD camera, webcam, and mobile. A mobile device is also known as a handheld device. Therefore, each of the display techniques must have a different type of suitable approach for the interaction. In addition, there are four types of interface in AR, tangible interface (TUI), Transitional Interface, collaborative AR interface, and hybrid. By using a collaborative AR interface, the AR view becomes collaborative by connecting between two or more users through the internet connection [7].

Mobile AR has ubiquitous availability, strong computing power and by relating physical-world objects and virtual information makes the mobile AR can support the situated or immersive perception of a complex concept [9]. Besides that, interaction is one of the critical issues in AR in terms of accuracy and robustness. It is because implementing an appropriate interaction technique can affect the user experience in AR. The 3D object manipulation has been a challenge for the interface designer since it involves the control of six degrees of freedom (6DOF) [10]. Furthermore, to improve the communication between the user in AR is by making the AR become collaborative. Collaboration in AR can improve the users' awareness and give attention to their surroundings with the local user [8].

2. Handheld AR Application

The handheld device is used as a display technique. Before it can be used, a few configurations need to be made. The right API is required to run the application on the handheld device without any problem or error. Besides that, the handheld device also comes with a built-in camera. The camera is used to video stream the target marker. If the target marker is recognized as a registered marker, then the AR 3D Model appeared on the device display screen at the center of the marker. Furthermore, the handheld device also comes with a touchscreen screen.

2.1. Methodology

The first phase describes the user interaction within the collaborative AR interface environment. It is crucial to identify the remaining issues in handheld AR and the fundamental of AR. It also includes the touchscreen interaction for the handheld device. The findings of this phase have been further discussed in the literature review.

The second phase is to develop an AR application in a handheld device and to design AR collaborative interface. At this phase, the logic is being coded, and the AR view is enabled. After that, user interaction with more than one user, a multi-user interaction has been created to actualize collaborative AR interface Multi-user interaction applied touch gesture for interaction and to allow collaborative works, an internet protocol is determined. In order to enable AR technology in a handheld device, a natural feature tracking has been executed in this phase. ARGarden application has been developed with the capability to add more than one user to interact with the 3D contents.

The third phase is to integrate ARGarden application with multi-user interaction for collaborative AR. At this phase, photon service is used to enable two or more users to connect with each other during collaborative AR. Figure 1 shows the methodology of this research.

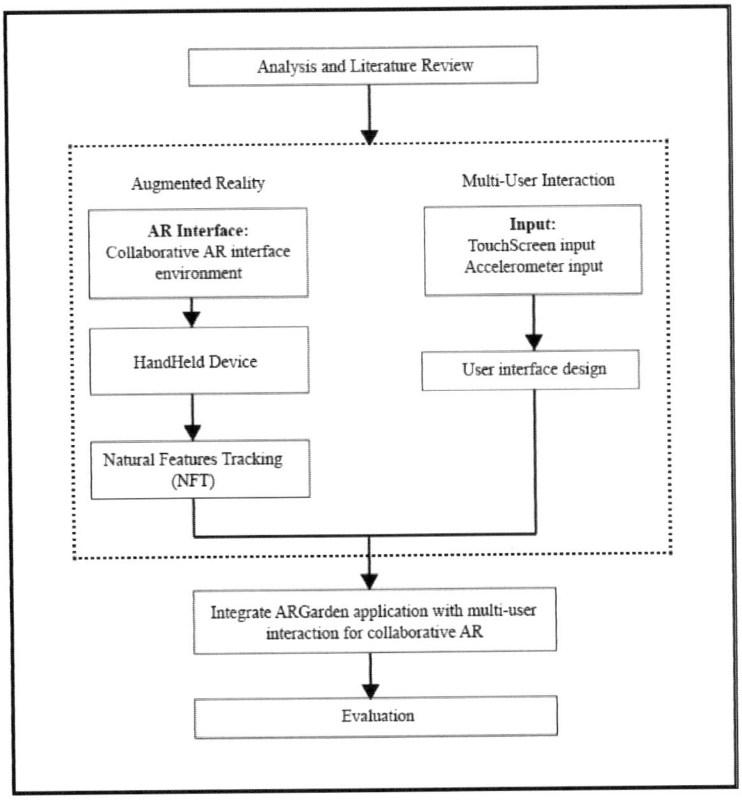

Figure 1. Research Methodology.

2.2. *AR Handheld Interface*

The features-based is a tracking technique that detects the feature on a real image by using a camera. Figure 2 shows our RGB marker has been converted into features. We used a single marker, coloured image in RGB as our image target and uploaded it into the database in the Vuforia SDK. This features-based tracking system will convert our image target from RGB to grayscale version to identify the features that can be used for recognition and tracking.

Figure 2 (a) explains the process of converting an RGB image into features. The marker tracking process has been used in this research; a printed image with RGB colour is tracked by a camera, then the image is converted into grayscale as in figure 2 (b), and with the default threshold values, the image has produced the features, as in figure 2 (c). The features are generated and stored in a dataset. The marker with the unique pattern is recognized by their features; once the camera captures the image, the dataset is retrieved. The features are used to register an object. Once the camera detects the image with the assigned object, the object is overlaid on the top of the image.

(a) (b) (c)

Figure 2. Natural Features Tracking (NFT) for AR Tracking.

Figure 3 presents the AR interface architecture for this research. The virtual element contains the data of the 3D object. This data is used to spawn the object in the AR scene in format. fbx. This 3D data has the 3D geometry and the texture of the object. Besides that, the tracking technologies used feature tracking, which is feature-based tracking. Besides that, the interaction used is multi-user interaction for a handheld device. The interaction that is involved in 3D object manipulation is positioning and orientation. The result from this interaction was calculated and displayed through the handheld display.

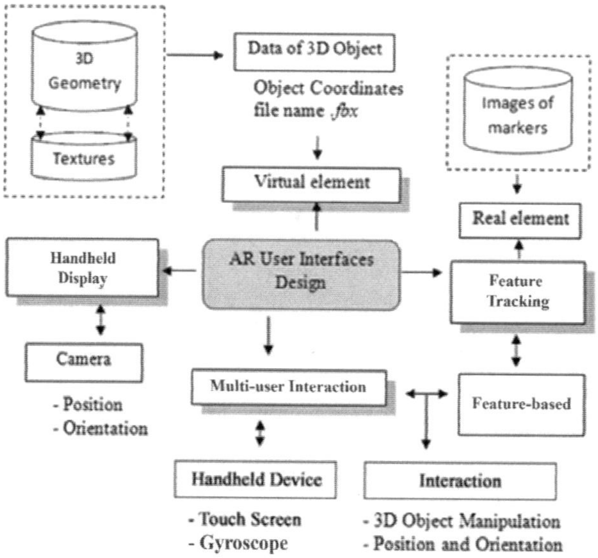

Figure 3. AR User Interface Design with Handheld.

2.3. *Touch-Screen and Gyroscope Input*

The input used in this research is touchscreen input and gyroscope input, as shown in Fig. 3. Before the device tilting input can be received, the object needs to be selected first by using touchscreen input. If the touch input receives, then the raycast is generated, and it checked if the raycast is touching any object. When the object is touch, it gives feedback to the user by highlighting the object and creating a line between the user device and the selected object. Then, any translation of the object can be done by swiping the touch input on the screen. Besides that, the accelerometer is a technology that is used to measure tilt angle, rotation, and it is applied in changing the device screen orientation vertically or horizontally. For this research, the accelerometer is used to manipulate the object rotation by tilting the device. After any translation has been done, the viewpoint is calculated before the result display on the screen.

Table 1 shows the touchscreen gesture that has been used to get input from the user. (a) Tap gestures are used to instantiate the object. Tap is a gesture by using one index finger to touch on the screen and the gesture to scale up and scale down the object by using (b) spread and (c) pinch. Spread is a gesture by spreading the thumb finger and index finger apart while touching on the screen index while the pinch is a gesture by pinching the thumb finger and index finger together while touching on the screen.

When the object is touch, selected or deselected, the touched object is sent through the network, and it tells all users the state of the object. If the user is the owner of the object, then the object can be manipulated by the user to do the transformation. Table 1 shows the tilting gesture when doing the rotation transformation. When the user tilts the device to the right (a), the object is rotated to the right, and when the user rotated to the left, as shown in (b), the object is turned to the left. The tilt input is received from gyroscope input.

Table 1. Interaction using touchscreen and gyroscope.

Touch-screen gesture			Tilting gesture	
			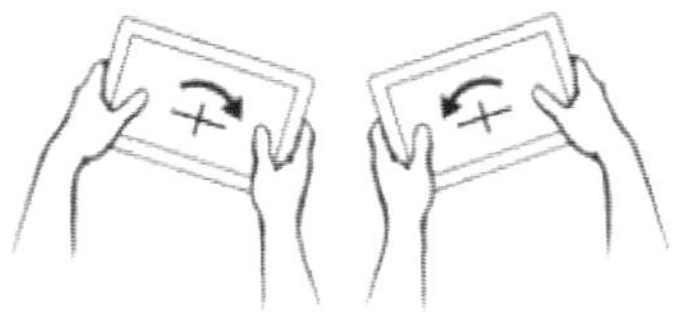	
Tap	Spread	Pinch		
(a)	(b)	(c)	(a)	(b)

3. Multi-user Interaction

The handheld device also comes with a built-in camera. This camera is used to video stream the target marker. If the target marker is recognized as a registered marker, then the AR 3D Model appeared on the device display screen at the centre of the marker. Furthermore, the handheld device also comes with a touchscreen screen. Input that is used in this research is touchscreen input and gyroscope input. Before the gyroscope input can be received, the object needs to be selected first by using touchscreen input. If the touch input receives, then the raycast is generated, and it checked if the raycast is touching any object. When the object is touch, it gives feedback to the user by highlighting the object and creating a line between the user device and the selected object.

Before this stage, the application was working only for a single user. Since the AR need to be collaborative, the user needs to be connected with each other through the network connection. In order to enable this environment, Photon Unity Networking (PUN) services were used, and figure 4 shows the flow process for multi-user.

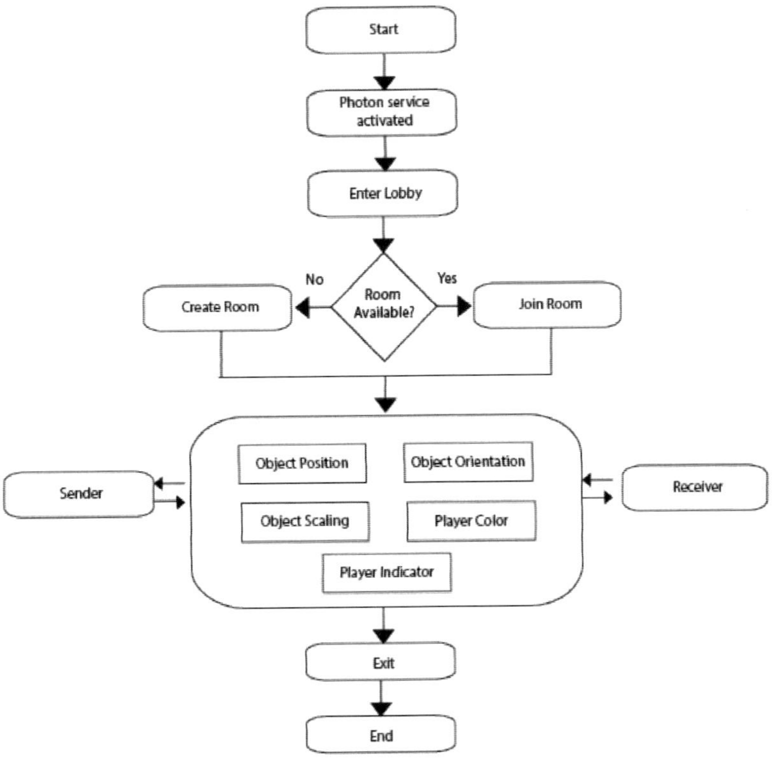

Figure 4. Multi-user Interaction in AR.

IOP Conf. Series: Materials Science and Engineering **979** (2020) 012001 doi:10.1088/1757-899X/979/1/012001

First, to activate and use the photon services, the photon app id needs to be created on its website, and this id is used for the photon configuration. After the service is activated, the user can enter the lobby. Then the user can create a room for the other user to join it. The user is responsible for creating the room is assigned as the master client. After the users connected in a room, process sender and receiver happened through the transmission control protocol (TCP). During this process, the data that is sent and received by the user is the object position, object orientation, object scaling, player colour, and player indicator. The object position, orientation and scaling were manipulated by using touchscreen input and gyroscope input.

There are two types of selection which are single selection and multiple selections. The implementation of a single selection is shown in figure 5 (a), and multiple selections are shown in figure 5 (b). Touch-screen and gyroscope input were used to enable the user to interact with the object in the AR environment. The user can move the object by dragging their hand on the screen. The gyroscope is used for the user to rotate the object while touching the screen. The indicator is spawned when the object was selected as in figure 5 (c). This indicator also can be seen by the other user in blue colour.

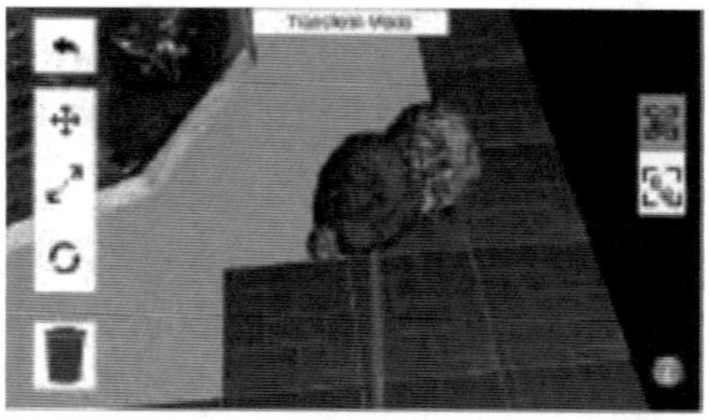

(a)

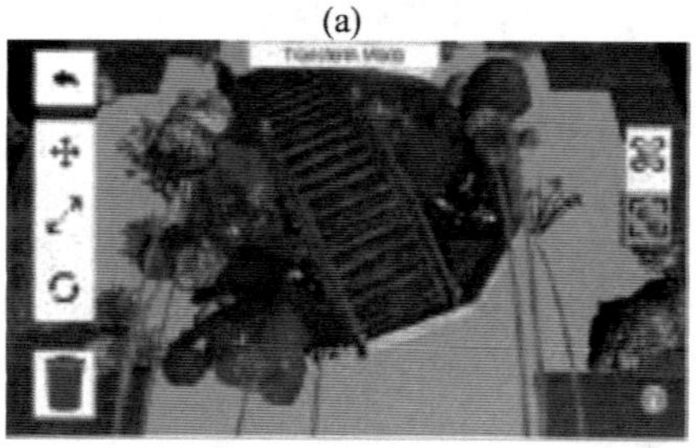

(b)

(c)

Figure 5. (a) single selection, (b) multiple selection, (c) multi-user selection.

Next, a few transformations have been applied, which are translation, rotation, and scaling. Figure 6 (a) shows the translation and figure 6 (b) shows the rotation. Figure 6 (c) shows the scaling and figure 6 (d) shows the uniform scaling, to perform scaling for the multiple object selection.

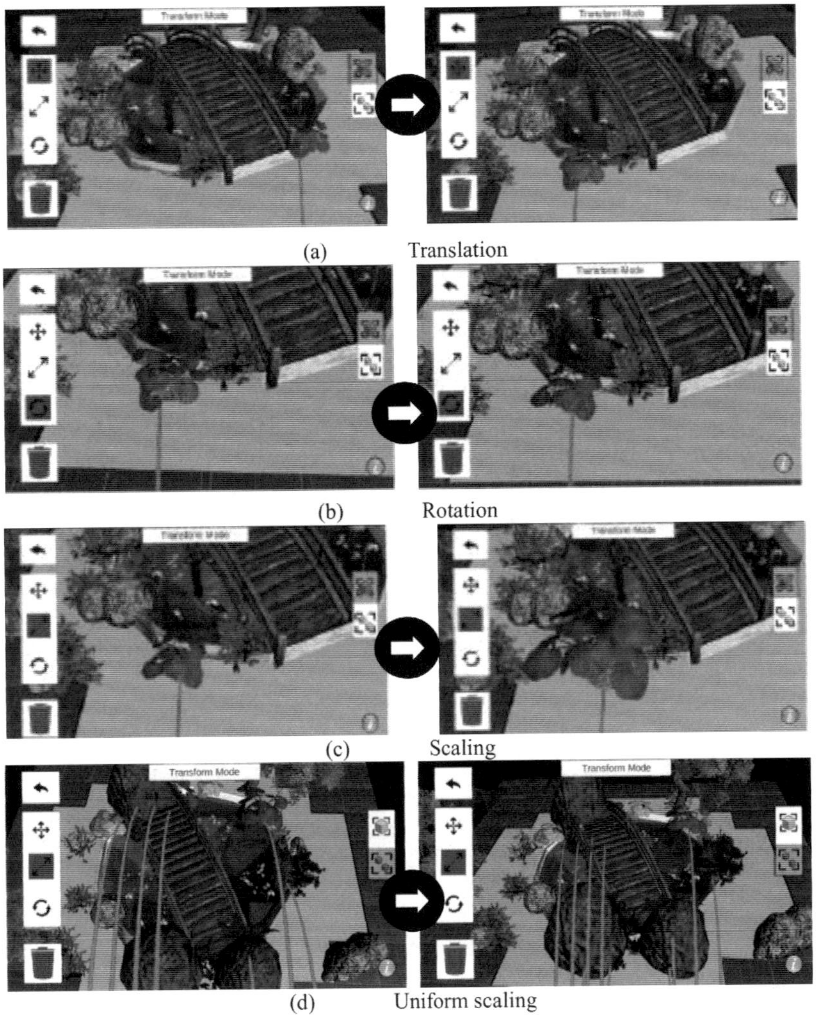

Figure 6. Manipulation in ARGarden application.

4. 3D Outdoor Landscape Design

ARGarden application is a prototype to implement method multi-user interaction for collaborative handheld AR. In this application, there are a few features and options that have been implemented. First are the features-based tracking technique. Even a part of the marker has been occluded, and the AR still can be displayed because another part of the marker can still be recognized by using feature-based tracking. In addition, in Vuforia, they provided a feature called extended tracking. This feature has been implemented in this application. Extended tracking is a concept that enables the target pose information to be still available even the target is no longer in the camera field of view. Therefore, by using this feature, it can improve the tracking performance and continue tracking even the target no longer in view.

Next, the network service is used in this application to create AR collaborative environment. This environment every object in the scene to be synchronized by the server. When it is involved multiple users, a queuing system has been implemented. Therefore, by using the queuing system, the user cannot touch or select any object that is currently selected by the other user. The user was also given a notification to tell them that they are already connected to the lobby. Once the users enter the application, the first screen has appeared as in figure 7 (a). The user was also provided with a tutorial when the first time they were using the application. The transform mode interface has appeared when `EditWorld` button is clicked. `Camera` button is used to take the screen-shot of the application, and UI is disabled during the screen-shot is taken. The screen-shot picture is saved in format .png. `ItemList` button allows the user to choose any type of object that they want to instantiate. `Information` button is used to display the tutorial as a guide for the user. Figure 7 (b) shows three users log-in into ARGarden.

(a)

(b)

Figure 7. Multi-user Interaction in ARGarden Handheld AR.

Once `EditWorld` is selected, the user is able to modify the 3D outdoor landscape. The `Back` button is used to allow the user to back to the main menu. The `Deselect` button also deselects all currently selected objects when transform mode changes to the main menu. The `Translate` button allows the user to translate the selected object and the `Scale` button allows the user to scale up or scale down the selected object. The `Rotate` button allows the user to rotate the selected object, and the Select button allows the user to select only one object at one time. Lastly, `Multi` button allows the user to select more than one object at one time, and the `Delete` button allows the user to delete the selected object. All available buttons have been assigned as in figure 8.

Figure 8. Transform Mode in `EditWorld` option.

5. Evaluation

In the experiment setup, the physical printed A3 marker was placed on the table where the users can move around the table to scan the marker. Two users with the handheld device are needed to create a collaborative multi-user interaction environment. Each of the handheld device used has a built-in camera that is used to track the marker, and the devices have a feature to receive the touchscreen input for the user to interact with the virtual object. Then, all the handheld device used also has an internet connection that is used to connect the users by using the network.

The participant age is between 22 and 23 years old, with a total of 10 students. They were given an instruction manual on how to use the prototype before working on their own. Then they were asked to fill the information form that asking them if they have any experience with AR and in using AR that is more than one user. Then, they were given time at least 4 min to interact with the 3D virtual object in the collaborative AR environment. Then the questionnaire and respondents' comments were used to collect the data. The pre-questionnaire asked the respondent about their experience in using AR and their experience in using AR with multiple users. Post-questionnaire asked the respondent about the feedback upon using this application.

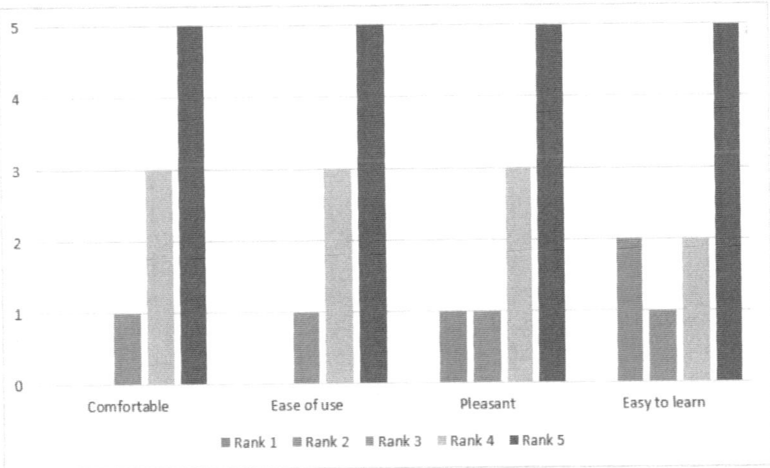

Figure 9. User experience with ARGarden.

The evaluation was guided referred to ARPad [14]. Then the questionnaire and respondents' comments were used to collect the data. The rank provides 5 levels, 1 indicates the lowest and 5 indicates the highest. The evaluation of ARGarden is to measure user experience in terms of comfortable, ease of use, pleasant and easy to learn. Figure 9 shows in comfortable user to the ARGarden application, 6 respondents have assigned Rank 5 and 3 respondents assigned moderate with Rank 3. In terms of ease of use, the results show the same to the comfortable. The users feel pleasant with the application, and however, in terms of easy to learn, the users feel joining a room to invite multi-user is hard when the task interrupts with the internet connection. A stable network is required to work with the ARGarden application. User feedback is positive on the colour indicator as visual feedback to differentiate the users in ARGarden. For future work, there will be more evaluation needed to measure the application on how touchscreen and gyroscope are able to improve the user experience when using the application.

6. Conclusion

This paper has discussed in detail 3D outdoor landscape design using handheld AR with multi-user interaction implemented in ARGarden application. A few limitations, however, were identified in this research which one of it is the internet connection must be fully functional for the collaborative AR part to work. Without an internet connection, the users cannot connect with each other and the AR view cannot synchronize. Besides that, the object also cannot update its position correctly if the internet connection was unstable or slow.

As for future work, there are a few recommendations that can be applied to improve the user experience when using the application. First, the marker-based tracking can be changed into marker-less tracking. Therefore, the user can use the application anywhere without needing a physical printed marker. They can just scan a ground plane and set the area they want to display the AR. Besides, for the interaction, it was recommended to try to use the real hand gesture to enable the user to interact with the virtual object just like the real world.

Acknowledgements

We would like to express our appreciation to Mixed and Virtual Reality Laboratory (mivielab), ViCubeLab at Universiti Teknologi Malaysia (UTM) for the facilities and technical supports.

References

[1] L V Zhihan, L Xiaoming, and L Wenbin 2017 Virtual reality geographical interactive scene semantics research for immersive geography learning *Neurocomputing* 254 71-78

[2] J M Mota, I Ruiz-Rube, J M Dodero, and I Arnedillo-Sánchez 2018 Augmented reality mobile app development for all *Computers & Electrical Engineering* 65 250-260.

[3] S G Dacko 2017 Enabling smart retail settings via mobile augmented reality shopping apps *Technological Forecasting and Social Change* 124 243-256.

[4] T K Huang, C H Yang, Y H Hsieh, J C Wang, and C Hung 2018 Augmented reality (AR) and virtual reality (VR) applied in dentistry *The Kaohsiung journal of medical sciences* 34(4) pp 243-248.

[5] F E Fadzli, and A W Ismail 2019 VoxAR: 3D Modelling Editor Using Real Hands Gesture for Augmented Reality *In 2019 IEEE 7th Conf. on Systems, Process and Control (ICSPC)* pp 242-247

[6] A W Ismail, M Billinghurst, and M S Sunar 2015 Vision-Based Technique and Issues for Multimodal Interaction in Augmented Reality. *In Proc. of the 8th International Symp. on Visual Information Communication and Interaction* pp 75-82

[7] F Zhou, H B L Duh, and M Billinghurst 2008 Trends in augmented reality tracking, interaction and display: A review of ten years of ISMAR. *In Proc. of the 7th IEEE/ACM Int. Symp. on Mixed and Augmented Reality* pp 193-202

[8] D Datcu, S G Lukosch, and H K Lukosch 2014 A collaborative game to study the perception of presence during virtual co-location *In Proc. of the companion publication of the 17th ACM conf. on Computer supported cooperative work & social computing* 5-8

[9] F Ke, and Y C Hsu 2015 Mobile augmented-reality artifact creation as a component of mobile computer-supported collaborative learning *The Internet and Higher Education* 26 33-41.

IOP Conf. Series: Materials Science and Engineering **979** (2020) 012001 doi:10.1088/1757-899X/979/1/012001

[10] A W Ismail. *User Interaction Technique With 3D Object Manipulation in Augmented Reality Environment* 2011 (Doctoral dissertation, Universiti Teknologi Malaysia)

[11] A W Ismail, Noh Z Ismail, M S Sunar, and M Najib 2008 Augmented Reality Theory and Applications. *Advances in Computer Graphics and Virtual Environment* 88-105.

[12] P Milgram and H Colquhoun 1999 A taxonomy of real and virtual world display integration. Mixed reality: Merging real and virtual world (vo1, pp 1-26.)

[13] A W Ismail, and M S Sunar .2009 Collaborative augmented reality: multi-user interaction in urban simulation *In Int. Visual Informatics Conf.* pp 382-391

[14] D Mogilev, K Kiyokawa, M Billinghurst and J Pair 2002 Ar pad: An interface for face-to-face ar collaboration *In CHI'02 extended abstracts on Human factors in computing systems* 654-655

Bio-WTiP: Biology lesson in handheld augmented reality application using tangible interaction

C S Yusof [1,2]*, N L S Amri [1,2] and A W Ismail [1,2]

[1] Mixed and Virtual Environment Research Lab (mivielab), ViCubeLab, Universiti Teknologi Malaysia, 81310 Johor, Malaysia
[2] School of Computing, Faculty of Engineering, Universiti Teknologi Malaysia, 81310 Johor, Malaysia

* suhaimi@utm.my

Abstract. Augmented Reality (AR) is a combination of objects of both real and virtual that partially merged into a single environment. In the education field, the process of teaching and learning can be more interesting with the engagement of AR as aided learning tools during classes. However, the remaining problems are, Biology was one of the subjects that hard for students to easily understand such as learning water transport in the plant. The use of touchscreen in handheld interaction is limited due to screen size for user to use in handheld AR interaction for 3D object manipulation. Therefore, this project aims to develop a Biology lesson handheld AR application using tangible interaction (Bio-WTiP AR). There are three phases that have been carried out, which are AR feature-based tracking, integrate Biology application with tangible interaction in handheld AR and evaluation. The testing and evaluation have been performed based on usability and user acceptance. The results show the application has been accepted by the users as the users found it is interesting to learn the process of water transport in the plant on hand. Based on the results, this project has successfully produced a Bio-WTiP application that can enhance the users learning process at once gave users experience in AR.

1. Introduction

Augmented Reality (AR) is an upgrade version of reality where life directly or indirectly reads of physical actual-world surroundings are jumbled with layered computer-generated images over user view of the actual-world [1]. Nowadays, teachers either in primary or secondary school prefer to use digital technology as a medium for them to deliver their lessons such as slide notes, videos, or three-dimensional (3D) graphical multimedia. Hence, AR is one of the newest technologies that enable seamless connection in combining physical and virtual worlds that have applied to education.

This paper presents the development of Biology lesson handheld AR application using tangible interaction, namely Bio-WTiP. It is for science stream student in secondary school with the implementation of AR technology in it. Instead of learning just by using textbook which is too ordinary, this application will provide better learning concept, and 3D graphic as students also can enjoy and experience the new growing technologies in the application. Since we are living in the era that encountering a fast penetration of smartphones, portable devices, and mobile application, it makes sense for the student to start learning using these gadgets instead of books as now all information just one touch away.

The Bio-WTiP application allows the student to manipulate with the displayed 3D graphics using tangible interaction. By using tangible interaction, the student can take control of the model display by adding water molecule to the plant as this project focuses on water transport in the plant. They need to hover the mobile phone's camera to the marker, which is the plant marker provided for the app to displays the 3D graphics and student can interact with it. Therefore, this paper describes the literature review, development process and the Bio-WTiP application. It ends with the conclusion after discussing the findings.

2. Literature Review

2.1. Augmented Reality

AR is a system that seamlessly permitted users to see the actual-world environment with the virtual objects overlaid upon it. There are three major facets in the AR system that fuses real and virtual objects in the actual-world environment. Which are, real-time, runs interactively, and roster real and virtual objects with each other [2]. The definition of Milgram's Reality-Virtuality Continuum was discovered by Paul Milgram and Fumio Kishino. It is an interval between the actual-world environment and the virtual environment, in between as a continuum that, where AR is closer to the actual world and Augmented Virtuality (AV) is close to a real virtual environment. AR also relies on the registration, such as alignment of the digital content, which is "augmentation" on the physical surrounding (digital registered marker) [3].

2.2. Feature-Based Tracking

Nowadays, most people prefer to use feature-based in their AR application instead of a marker-based system. To registering virtual objects into the actual-world environment, software development kit (SDK) like Vuforia is needed as a platform for AR applications. Vuforia, developed by Qualcomm, allows different AR applications to be produced and helps to recognise and monitor planar images and 3D objects in real time [4]. By this platform, it would roster and tracks the objects of the actual world and matches them with objects of the virtual world. Then the 3D model is controlled through Rajawali [5]. The key features in Vuforia incorporate the capability of the software to recognize and track images, objects, text, markers and modernize environments [6]. According to [4], there are no repeated patterns present in the image goal to achieve a better image recognition efficiency, and they must have high local contrast, uniform distribution of features and high density of features. This SDK supports the build-in Android, iOS and Universal Windows Platform (UWP) applications for mobile applications.

2.3. Tangible User Interface (TUI)

Handheld Augmented Reality (HAR) applications allow interaction with the physical environment of the user through tablets or smartphones on top of the real world with virtual content [7]. In general, handheld interaction device are widely understood as illusion lenses as users could use a camera on the opposite side of the screen to see the real world with simulated content through a monitor on the devices [8]. To manipulate the virtual objects in HAR applications, there are many interaction techniques that have been used by the user. AR handheld interaction has been categorized into two technique which is tangible interaction and intangible interaction [9] [10].

According to [11], the goal of TUI is to transform real objects into computer interface input and output devices. Originally, Ishii and Ullmer introduced TUI to describe novel human-computer interaction (HCI) types focused on the physical manipulation of everyday objects [12]. In term of interface, handheld AR interface can be advanced to Mixed Reality (MR) interface where user can use MR apparatuses to work with simulation in more immersive [13]. The tangible approach allowed intuitive interfaces to be developed, promoting collaborative activities, two-handed interactions, and providing a robust sense of straightness to users. [12] aforesaid that it is believed that contact with tangible interfaces is more normal or familiar than with other forms of interfaces and could be more available to young children, people with learning disabilities or novices, minimising the brink of participation. According to [14], tangible interaction provides various DOF for the user to interact with the immersive environment.

Paddle-based interaction is one of the tangible techniques that is relevant and efficient for use in education which has been studied by [15] and supported by [11]. Paddle-based interaction provides a choice on how to use a TUI in AR for virtual object scene assembly as it enables virtual objects to be controlled between a scene. The paddle is the key instrument for interaction device and is intended to be used by either hand to create static and dynamic behaviours. [15] also believed that it is very natural to conduct these forms of interaction with a real paddle. Thus, the user can easily use this method to apply tangible interaction in AR environment as its able user assemble a difficult arrangement of virtual objects in the AR applications.

Touch-based is one of the techniques in handheld AR for object manipulation [16]. The most common used touch-based input is via a touchscreen device. There a several issues regarding the use of touch-based as input to the AR application. Yusof et al. [17] has stated part of the issues regarding to AR experiences due a natural intuitive interaction technique instead of simply pointing and clicking, and manipulating 3D content from the 2D gesture. In addition, Bai et al. [18] supported the statement from [19], it is due to small size device the interaction area and display is limited which caused the limitation in occlusion of the finger on the screen content. This issue caused the user to feel uncomfortable to play around with the application freely. Thus, the use touchscreen in handheld interaction is too usual for the user thus tangible interaction is used in handheld AR interaction [20] and [21] stated that current interaction method mainly finite to clicking and 2D touch.

Besides that, hand gesture interaction also is one of the inputs that allow users to interact in more naturally and intuitively [22]. Gesture-based interaction interprets the user's, actions separated from the device [23]. Usually, to allow a real-time hand gesture interaction, an external device is needed to combine the gesture in the context of AR environment with the physical world such as Leap Motion Controller. The device used infrared (IR) to determine the position of the objects in a limited real time-space [22]. The use of an external device like Leap Motion Controller required a lot of budgets and it is unsuitable to be practiced with a tablet as Leap Motion required to connect via USB to a computer [24].

There are drawbacks for TUI, however, as it can be vital to modify certain physical properties, making it difficult to inform by observing what the state of the digital data associated with that object is at a physical object. Therefore, the limitations can be overcome through the employment of AR. Besides using a handheld device, the tangible interaction also can be extended into using bare hands to interact [22].

2.4. AR in Education

There are multiple user learning environments discussed in [25] that, through the adaptation of Virtual Reality (VR), AR and MR, can create an immersive educational experience. By implying mobile AR application in education, it can enhance the students' learning understanding capacities [26]. The learning process of certain subjects also can be more interesting by using AR [7]. Usually, in educational institutions, lecture notes are famous teaching materials. However, educators are always looking for a new way to teach students and exploring new technologies that can give advantages to students in learning efficiently. Thus, it leads to found that AR is one of the latest technologies that considered as having a potential for pedagogical applications [27]. Furthermore, [28] has claimed that AR has a great potential to give out the valuable context for education as it allows learning method with experiences that can connect the information from the real world.

3. Related Works

Immersive authoring of tangible AR application is a system that suggested new approached for authoring tangible AR proposed by [29] that been called immersive authoring. Tangible AR approached which mixed interface of tangible user input methods with the output of AR display. This approach was suggested for virtual content in AR interface so it can be employed using a physical object. The main purposed of this approach also to created intuitive AR interface. It depended on precise tracking of the real object while using tangible AR interface and to achieve this requirement, ARToolKit computer vision tracking library was used. This is because the direction and orientation of the camera relative to square fiducial markers could be determined by ARToolKit, making it easier to create a tangible AR application [29]. Figure 1 below shows the prototype that was developed by [29] where it shows the virtual objects before interacted with each other (Figure 1 (a)) and when both markers interacted to one another (Figure 1 (b)).

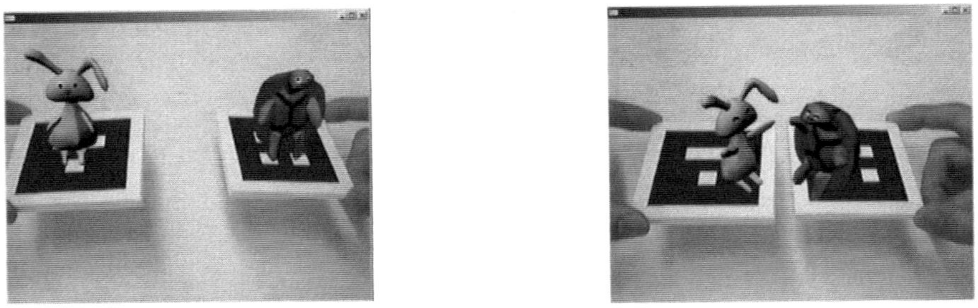

(a) Before interaction (b) After interaction

Figure 1. An interactive Tangible AR application [17].

SHARED SPACE Space-Multiplexed Interaction in AR was one of prototype example developed by [30]. The application applied the idea of tangible AR, which was used to develop a simple but powerful spaced-multiplex AR interface for several users. As it was designed to build a multiplayer AR game that does not involve user training, it offered users to play the AR application with another user. Furthermore, SHARED SPACE Space-Multiplexed Interaction needed the user to wear an Olympus HMDs with the camera attached to it as a medium for users to view the 3D graphics in the virtual world. It also provides users with multiple cards to select and look at the models from any angle. It is a game for users to find the matched object. When both objects are matched, the animation was triggered. While this is only a basic application, it has given a great test of the utility of a tangible interface for virtual object manipulation [30].

Students usually have a hard time to understand terms in science or a process that involve in certain subtopics during classes. Some schools did not provide a projector in their classroom for teachers to show a video during teaching. This is because an AR Buku Teks application was developed by Xentral Methods in 2016 to easier the teaching process in class as it required a mobile phone for display. This AR application developed especially for Malaysian textbooks for primary and secondary schools. The application has covered syllabus for science Form 1 and 2 only. The application is to assisted students in better understanding the subject in hand via 3D modelling, videos, and audio. Users needed to scan the image target which is images from the textbook itself for displaying the 3D model as shown in Figure 2 which consisted of cell structure function (Figure 2 (a)) and earth's layers (Figure 2(b)).

(a) Cell structure function (b) Earth's layers

Figure 2. AR buku teks content.

4. Proposed Bio-WTiP AR Application

This section explains the system overview of the Water Transport in Plant for Biology lesson in handheld AR application using tangible interaction or in short, we named as Bio-WTiP AR application. The following subsection includes the 3D object modelling, 3D object animation and the project interface.

IOP Conf. Series: Materials Science and Engineering **979** (2020) 012002 doi:10.1088/1757-899X/979/1/012002

4.1. Tracking and Bio-WTiP Workspace

Tracking process implemented to display the virtual object on the top of the real world, and the use of Vuforia SDK in this project is to track the target images and allowing them to interact with space in the real world. In Vuforia, the way to track the target manager image was by using Vuforia target manager. Vuforia Target Manager is one of Vuforia library that has a web-based tool that enables developers to create and manage target databases online. The virtual object was displayed on the screen right after the camera recognized the features of the image target that has been registered in the Vuforia database. Figure 3 below shows the tracking process and user workspace.

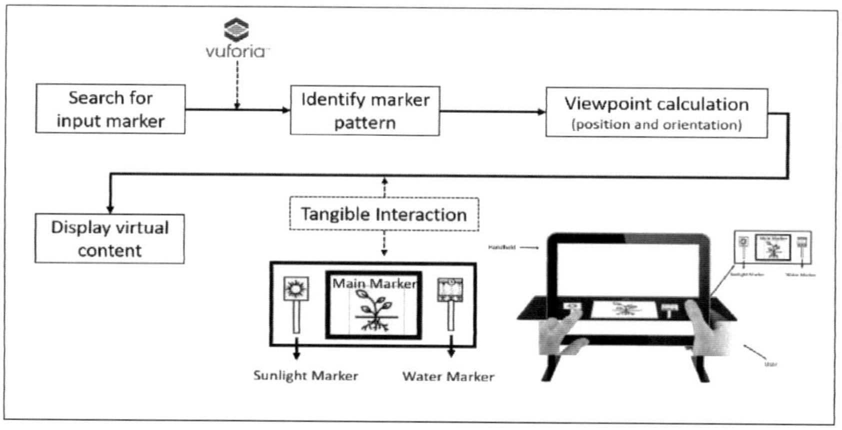

Figure 3. Tracking and AR workplace.

The tracking process that takes place when the user was required to test the project were displayed in Figure 3. At first, the AR camera searched for the input marker. This occurred when the user hovers the handheld's camera to the printed marker, which is the Plant marker. Then, the input marker must match with the pattern of the marker that has been uploaded in the Vuforia database. The viewpoint calculation allowed the process of recognition of the marker and allowed the virtual object appears on the top of the marker. Tangible interaction happens between one marker (Sunlight Marker) with another marker (Water Marker) that takes place until the display of the virtual content has appeared on the screen.

4.2. Tangible Markers

Tangible AR interface mixed the augment display potential of AR with intuitive interaction and manipulation of a physical object. It also supports the interaction between actual and virtual worlds as it provided with natural interaction range. This project utilized the used of tangible interaction between multiple markers to control the manipulation of the virtual content in the application and the use of virtual buttons to give out input to the user for better understanding. Figure 4 below shows all the marker used in this project, and each marker holds different virtual content and purposes.

The plant marker (Figure 4 (a)) allows the display of a 3D AR plant model in the scene while the sunlight marker (Figure 4 (b)) holds the sunlight constraint. The water marker (Figure 4 (c)) allowed the display of 3D AR water molecules in the scene between the stem of the plant, and lastly, the virtual buttons marker (Figure 4 (d)) holds multiple virtual buttons where users interacted with it to give out more information about the learning process. Three processes occurred when the markers interact with each other. The 3D model plant needed to be presented throughout the process for users to manipulate the virtual content displayed.

(a) Plant Marker (b) Sunlight Marker (c) Water Marker (d) Virtual Buttons Marker

Figure 4. List of markers.

The expected number of markers used in this project was four in total. First is the main marker, the other two is the elements or constraints to the AR environment which is, water molecule and presence of sunlight, and lastly is a virtual buttons marker that contained five buttons elements.

4.3. Simulation Data

The data of 3D models in the project are modelled using Blender 2.8 as it was easier the process to shape the primitive shape such as cylinder to the Bezier curve for the body of the plant. When all the shapes were joined to create a plant, the materials for the model is set. Then the 3D model will be exported into FilmBox (FBX) file format. The FBX file format is a well-known 3D data interchange format utilized between 3D editor and game engines. As the 3D model need to be imported into the Unity, FBX offers the right choice as it can support a lot of combined data as the 3D model need to keep the 3D bones models and the colour materials of the model. All the process mentioned illustrated in Figure 5 below.

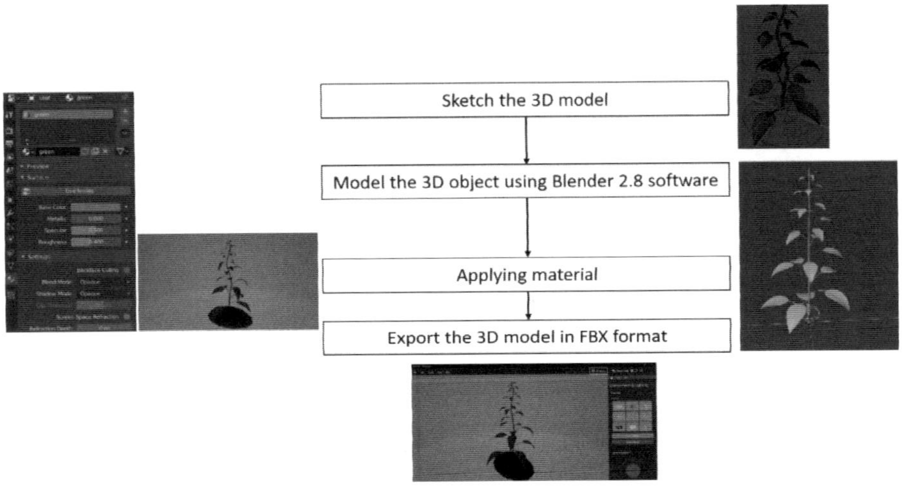

Figure 5. 3D Modelling process for 3D object.

The 3D plant model that has been designed and used in this application is shown in Figure 6. Figure 6 (a) shows the basic plant model without the adding of materials to the surface of the 3D model plant, and Figure 6 (b) shows the plant model with adding of material. There are three parts needed to be ready in creating the plant model, which are the base of the plant which represent the ground with soil, the stem of the plant and the leaf of the plant. These three parts have been created separately and joined at the end of the process. Curve modifier was added at the stem and leaf of the plant in Blender 2.8 to give a curve shape to the plant model. It is to make the plant looked real in user views instead of a basic solid straight plant model. To avoid a holographic viewing on the plant's stem when the plant was sliced, select the faces of the plant's stem in Blender edit mode, then click *Shift+N* to add a normal line to the surface of the plant model

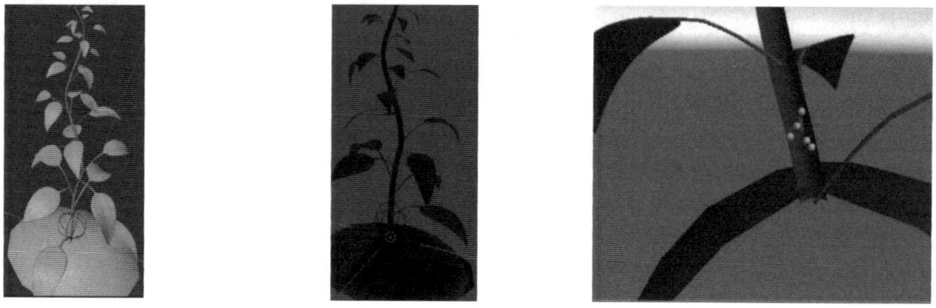

(a) Basic plant model (b) Model with material (c) A model with material and addition of animation

Figure 6. Plant Model.

The animation was created by adding a new Animator Controller than the animation clip was recorded by controlled and placed the position of x, y, and z of the model by the keyframes in the timeline. It easier as the position flow of the water molecule can be controlled to the exact position in between the plant's stem model. Figure 6 (c) shows the animation of the water molecule in between the stem of the plant run in Unity 3D. There are several water molecules has been added into the plant's stem where the molecules animated once it has spawned into the environment.

5. Implementing Tangible Interaction in Bio-WTiP

As presented in Figure 7, the flowchart to demonstrate the object manipulation using tangible interaction in handheld AR. The user starts with the opening the camera, followed by scanning the targeted marker. After that, a virtual object which in this case is a plant augmented onto the marker through this application, and the user can also manipulate the water flow in the plant by adding another marker to the AR space. Besides, the user also can see an animation flow of water molecule transport from root until the leaf of the plant by click one of the virtual buttons displayed. When the user adding the water, there will be an audio of information and process flow regarding the current condition of the plant. User can play with the process of water transport in the plant by adding a water molecule once at a time and increase or decrease the intensity of the sunlight into the AR environment.

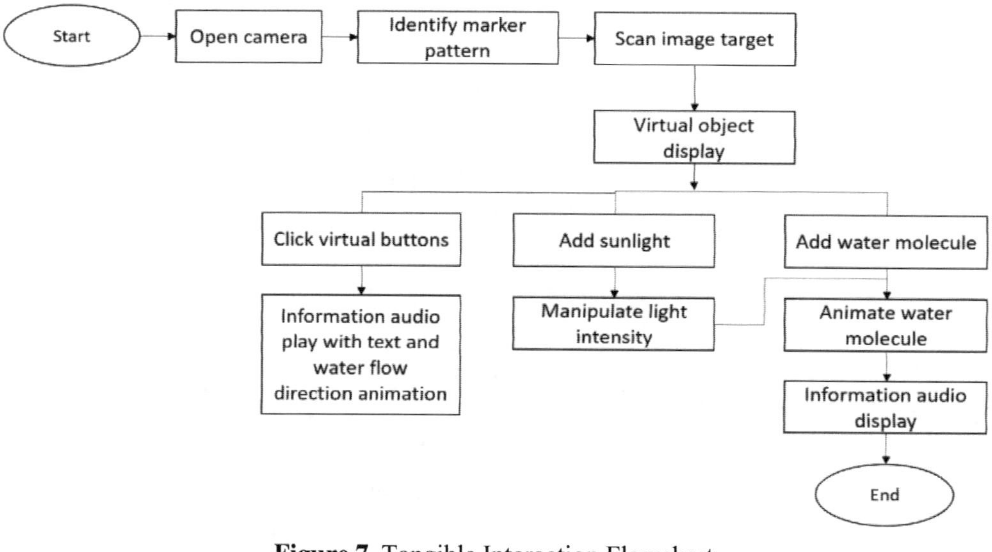

Figure 7. Tangible Interaction Flowchart.

Tangible interaction began as a user needed to add water molecules to the environment of the sliced plant model. The process of adding water molecules to the AR environment began when the camera detected the water marker. The water molecule spawned in between the plant's stem on the plant marker. The animation of water flows from the bottom of the plant until the leaf occurred. User can freely add the amount of water molecule to the plant, and each time the camera detects the water marker, a water molecule spawned to the environment.

Furthermore, the user can control the light intensity in the environment to see the change of the water molecule flow's speed. The light intensity of the environment can be control by using sunlight marker. The nearer the sunlight marker to the plant marker, the higher the value of light intensity which reflected the increasing of the water molecule flow's speed. Each marker used in this application held different object or purpose to it. Plant marker holds the 3D model of the plant. Water marker holds water molecule, an attached of note, and the trigger event to spawn the water molecule to the plant environment. The sunlight marker holds the sunlight of the environment and the trigger event to calculate the distance of markers to control the light intensity value.

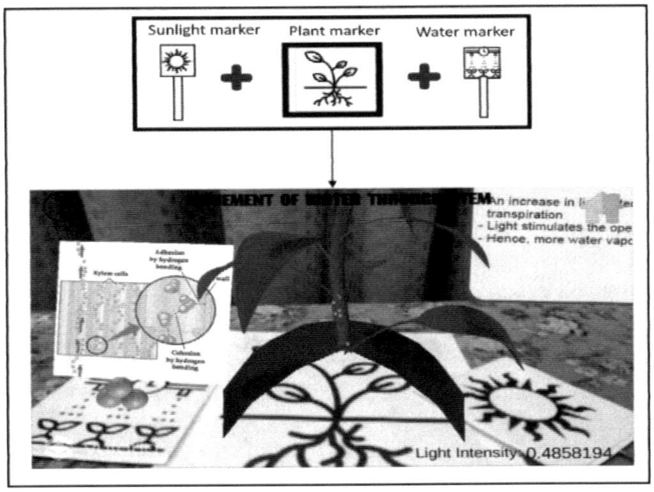

Figure 8. Tangible Interaction between markers.

Based on Figure 8, it shows the interaction that occurs when a water marker and sunlight marker detected by the camera. The water marker caused the spawning of the water molecule to the plant, and the sunlight marker increased the light intensity in the environment. Along with three of these markers, there are virtual buttons marker that user could use during the learning process. The virtual buttons marker consisted of five virtual buttons in totals, and all the buttons held different audio and text to it for the user to listen and read the information.

The virtual buttons are shown in Figure 9 (a) below. For "Info 4" button in Figure 9 (b), there is an animation occurred in the AR plant environment where there is an arrow indication showed the user the direction flow of the water transport from the root to the leaf, and the result is as shown in Figure 9.

(a) AR virtual button (b) Direction water flow

Figure 9. Virtual buttons with animation of water flow.

6. Evaluation

This phase was tested and evaluated the AR Biology lesson for mobile application using tangible interaction. It also focused more on user testing the prototype development and, the feedback from the test was collected to make further improvement if required. The test was conducted by Form 4 and Form 5 students who enrol in science courses as the application focused on Biology lessons.

The evaluation from the user testing process explained based on the bar charts. There are seven female and three male respondents to this application. Seven on the respondents did not have any experience in encountering AR technologies and three from the respondents have experienced AR technologies. The questions measured in a scale from range 1 (negative attributes) to range 5 (positive attributes) asked about respondents experienced and satisfaction in using the application. Figure 11 shows the evaluation results for Biology-Water Transport in Plant Augmented Reality (Bio-WTiP AR) application.

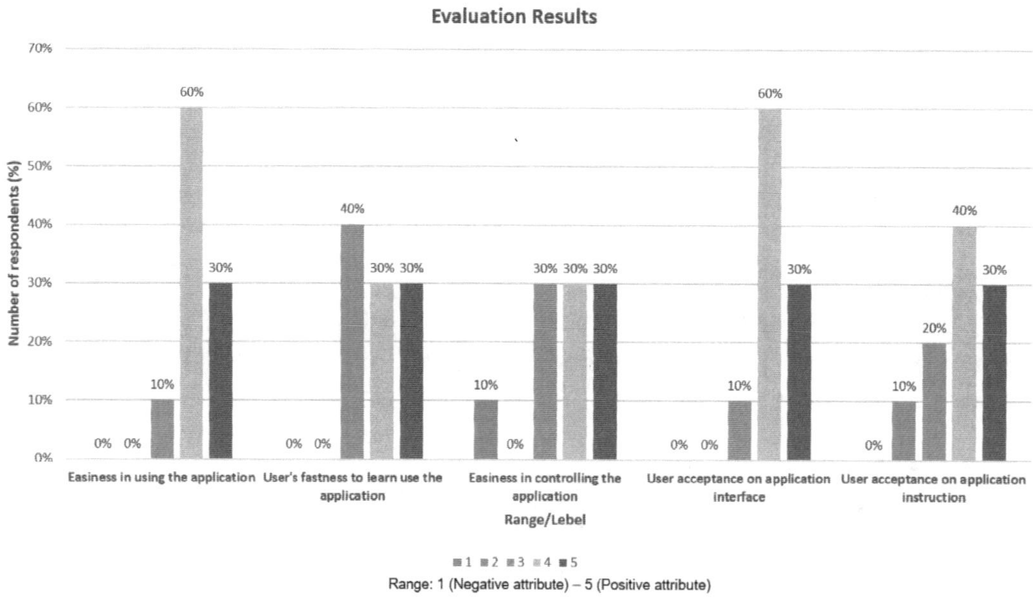

Figure 10. Usability Testing Results.

Figure 10 shows the results in bar charts of users respond to usability testing. For respondent's easiness in using the Bio-WTiP AR application, it clearly shows that most of the respondents found it quite easy to use the application with a percentage of 60% in total that agreed and 30% strongly agreed that the application is easy to use and understandable. Only one respondent which hold 10% of the results ranged the application on the third level which is not too hard and not too easy to use.

Next, respondent's fastness learned to use the Bio-WTiP AR application. Majority of the respondents found the application was fast to learn as 30% of the respondents range their level at the fourth, which represented as fast. Another 30% of the respondents range their level at the fifth, which represented as very fast. 40% of the respondents found the application was not too fast and not too slow to be learned, and none of the respondents is slow in pace to learn how to use the application.

Third, is the level of easiness for the respondents to control the Bio-WTiP AR application during the testing process. There are three levels that have the same number of respondents where it is range from in the middle of easy and hard, easy, and very easy. All these three levels held the percentage of 30% each of respondents and only 10% of the respondent found that the application is very difficult to be controlled during the testing process with none of the respondents ranged at the very hard and hard level.

Then, for respondent acceptance on the Bio-WTiP AR application interface, the graph shows that more than half of the respondents agreed that the interface design of the application was well designed. 60% and 30% of the respondent agreed and strongly agreed that the interface design was accepted to use. 10% of the percentage with represent of one respondent rated the design interface of the application at 3/5. Interface design is essential as it attracted the respondents to feel excited when interacting with the application and this shows that the design interface of the application was well accepted.

The fifth is the respondent acceptance on the clearer of the instruction in the Bio-WTiP AR application that given to the respondents. Based on the result, there are various answers to the question where 70% of the respondents' rate the application provide clear instruction to its users as they agreed (4/5) and strongly agreed (5/5) to the question given. 20% of the respondents rated 3/5 for the level as they found it is not too easy to understand the instruction given in the application and 10% of the respondents rated it is hard (2/5) for them to grasp the instruction provided.

Overall, all of the respondents were satisfied with the Bio-WTiP AR application as 50% of respondents rated 4/5 that they are satisfied enough on the Bio-WTiP AR application and the rest is strongly satisfied with the application with percentage of 50% in total. The high satisfaction of the application came from the easiness of the application to use, the easiness to control the application and interaction in the application, how clear the instruction is given to use the application and how well the design of the application interface.

7. Conclusion

From the result, this project developed to easier students in remembering and understanding on how the process of water transport in plant's stem occur. The topic was picked because the student had a hard time to visualize the process that occurs in a plant as it is visible towards naked eyes and they tend to have a misconception towards water transport in plant process.

The project is limited to one external condition that affects the rate of transpiration which is the presence of the sunlight in the environment. The target markers used was already being fixed as it has stored in the Vuforia database cloud. The build of the Bio-WTiP AR application also is limited to Android users only as the application was built in android package kit (APK) file format. The interaction that been applied in the application is tangible interaction where user needs to interact with another target marker to interaction with the virtual object. Furthermore, the interaction was limited to adding of the water molecule in the plant, manipulation of light intensity and play of voice information from virtual buttons only.

For the future works, to have a better understanding of the lesson, the system should have a more external condition that affects the rate of transpiration which are temperature, humidity, and air movement. Based on the usability testing, the respondents have recommended improving the application in the dual language, in English and Malay. It can give the users a better understanding to align with their syllabus in school. Multi-user interaction is another suggestion for future work, where a collaborative interface can be leveraged across learning environment.

Acknowledgement
We would like to express our appreciation to ViCubeLab at Universiti Teknologi Malaysia (UTM) for the facilities and technical supports. Special thanks to UTMER Grants (Q.J130000.3851.19J22) for the funding. A big gratitude is dedicated to the parties that contributed the fundamental knowledge in this paper.

References

[1] Ismail, A.W., Billinghurst, M., Sunar, M.S. and Yusof, C.S. (2018). Designing an Augmented Reality Multimodal Interface for 6DOF Manipulation Techniques, in Proceedings of SAI Intelligent Systems Conference (pp. 309-322). Springer, Cham.
[2] Ismail, A.W., Billinghurst, M. and Sunar, M.S. (2015). Vision-Based technique and issues for multimodal interaction in augmented reality. ACM International Conference Proceeding Series., (pp. 75-82).
[3] Carmigniani, J., Furht, B., Anisetti, M., Ceravolo, P., Damiani, E. and Ivkovic, M. (2011). Augmented reality technologies, systems and applications. Multimedia Tools and Applications.
[4] Upadhyay, G.K., Aggarwal, D., Bansal, A. and Bhola, G. (2020). Augmented Reality and Machine Learning based Product Identification in Retail using Vuforia and MobileNets. Proceedings of the 5th International Conference on Inventive Computation Technologies, ICICT 2020., 479–485.
[5] Xiao, C. and Lifeng, Z. (2014). Implementation of mobile augmented reality based on Vuforia and Rawajali. In Proceedings of the IEEE International Conference on Software Engineering and Service Sciences, ICSESS.
[6] Amin, D. and Govilkar, S. (2015). Comparative Study of Augmented Reality Sdk's. International Journal on Computational Science & Applications. 5(1), 11–26.
[7] Pranoto, H. and Panggabean, F.M. (2019). Increase the interest in learning by implementing augmented reality: Case studies studying rail transportation. Procedia Computer Science. 157, 506–513.
[8] Gervautz, M., & Schmalstieg, D. (2012). Anywhere interfaces using handheld augmented reality. Computer.
[9] Gupta, A., Lin, B. R., Ji, S., Patel, A., & Vogel, D. (2020). Replicate and Reuse: Tangible Interaction Design for Digitally-Augmented Physical Media Objects, 1–12.
[10] Bai, H., Lee, G. A., & Billinghurst, M. (2012). Freeze view touch and finger gesture based interaction methods for handheld augmented reality interfaces. In Proceedings of the 27th Conference on Image and Vision Computing New Zealand - IVCNZ '12.
[11] Kawashima, T., Imamoto, K., Kato, H., Tachibana, K. and Billinghurst, M. (2001). Magic Paddle: A Tangible Augmented Reality Interface for Object Manipulation. Interface. 195, 194–195.
[12] Schneider, B., Jermann, P., Zufferey, G. and Dillenbourg, P. (2011). Benefits of a tangible interface for collaborative learning and interaction. IEEE Transactions on Learning Technologies. 4(3), 222–232.
[13] Aladin, M.Y.F. and Ismail, A.W. (2019). Designing User Interaction using Gesture and Speech for Mixed Reality Interface. International Journal of Innovative Computing. 9(2), 71–77.
[14] Saidi, H., Serrano, M., Irani, P., Hurter, C., Saidi, H., Serrano, M., Irani, P., Hurter, C. and On-body, E.D. (2019). On-Body Tangible Interaction: Using the Body to Support Tangible Manipulations for Immersive Environments, Springer International Publishing.
[15] Tuli, N., & Mantri, A. (2015). Interaction Techniques : helping user manipulate AR virtual content, (September 2016).
[16] Goh, E. S., Sunar, M. S., & Ismail, A. W. (2019). 3D object manipulation techniques in handheld mobile augmented reality interface: A review. IEEE Access, 7, 40581–40601.
[17] Yusof, C.S., Bai, H., Billinghurst, M., M.S.S. (2016). A Review Of 3D Gesture Interaction For Handheld Augmented Reality. 2(August), 15–20.
[18] Yin, J., Fu, C., Zhang, X. and Liu, T. (2019). Precise Target Selection Techniques in Handheld

Augmented Reality Interfaces. IEEE Access. 7, 17663–17674.
[19] Bai, H., Lee, G.A., Ramakrishnan, M. and Billinghurst, M. (2014). 3D gesture interaction for handheld augmented reality. In SIGGRAPH Asia 2014 Mobile Graphics and Interactive Applications on - SA '14.
[20] Chowdhury, S.A. (2017). Handheld Augmented Reality Interaction Technique. 10645(November 2016), 418-426.
[21] Bai, H., Gao, L., El-Sana, J. and Billinghurst, M. (2013). Markerless 3D gesture-based interaction for handheld Augmented Reality interfaces. In 2013 IEEE International Symposium on Mixed and Augmented Reality, ISMAR 2013.
[22] Fadzli, F.E. and Ismail, A.W. (2019). VoxAR: 3D modelling editor using real hands gesture for augmented reality. Proceeding - 2019 IEEE 7th Conference on Systems, Process and Control, ICSPC 2019. (December), 242–247.
[23] Bai, H., & Lee, G. (2012). Interaction methods for mobile augmented reality, p. 101.
[24] Daineko, Y., Ipalakova, M., Tsoy, D., Bolatov, Z., Baurzhan, Z., & Yelgondy, Y. (2020). Augmented and virtual reality for physics: Experience of Kazakhstan secondary educational institutions. Computer Applications in Engineering Education, 28(5), 1220–1231.
[25] Nor'a, M.N.A. and Ismail, A.W. (2019). A Review on Collaborative Learning Environment across Virtual and Augmented Reality Technology. IOP Conference Series: Materials Science and Engineering. 551(1).
[26] Sungkur, R.K., Panchoo, A. and Bhoyroo, N.K. (2016). Augmented reality, the future of contextual mobile learning. Interactive Technology and Smart Education.
[27] Nincarean, D., Alia, M.B., Halim, N.D.A. and Rahman, M.H.A. (2013). Mobile Augmented Reality: The Potential for Education. Procedia - Social and Behavioral Sciences.
[28] Iftene, A. and Trandabăt, D. (2018). Enhancing the Attractiveness of Learning through Augmented Reality. Procedia Computer Science. 126, 166–175.
[29] Lee, G.A., Nelles, C., Billinghurst, M. and Kim, G.J. (2004). Immersive authoring of tangible augmented reality applications. ISMAR 2004: Proceedings of the Third IEEE and ACM International Symposium on Mixed and Augmented Reality. (June 2014), 172–181.
[30] Billinghurst, M., Kato, H. and Poupyrev, I. (2008). Tangible augmented reality. Acm siggraph asia., 1–10.

IOP Conf. Series: Materials Science and Engineering **979** (2020) 012003 doi:10.1088/1757-899X/979/1/012003

A mobile game SDK for remote collaborative between two users in augmented and virtual reality

Y H Ong [1]*, A W Ismail [1], N A.Iahad [1], M R M Rameli [2], and R Dollah [2]

[1] School of Computing, Faculty of Engineering, Universiti Teknologi Malaysia, 81310 Johor, Malaysia
[2] School of Education, Faculty of Social Sciences and Humanities, Universiti Teknologi Malaysia, 81310 Johor, Malaysia

* ongyuehuei@gmail.com

Abstract. Remote collaboration is important in Augmented Reality (AR) and Virtual Reality (VR) that allow users to communicate and interact with each other. Mobile games are gaining its popularity due to its portability and lower in cost. Software Development Kit (SDK) allows developer to speed up the development process. However, most of the collaboration of these technologies only focus on either AR or VR independently. Besides, most AR and VR that had been developed are only available in expensive devices but not in mobile devices. Therefore, this research aims to create a remote collaboration between AR and VR with mobile games. There are four phases have been carried out, analysis phases, development phase, construction phase and evaluation phase. The evaluation has been performed and it is based on the usability and user acceptance. The usability results show the application is accepted by the user. User acceptance results show the application works accordingly. Based on the results, this research has successfully produced a working SDK for remote collaboration between AR and VR with mobile games.

1. Introduction

Augmented Reality (AR) and Virtual Reality (VR) technology are unique and powerful by itself. AR enhances the real world environment by sharing real and virtual world at the same time [1]. VR entirely replaces the real world by simulating a virtual environment [2]. Collaboration between AR and VR becomes an important aspect to exploit both technologies to the strength [3]. Development of technology and the decreasing market price of mobile devices added an advantage for these technologies to be supported in mobile devices [4]. Developing mobile game with technologies like AR and VR consumes time and required specific knowledge. The lack of interdisciplinary ideas of developing game, AR and VR increase the demand of related Software Development Kit (SDK) during implementation and design level [5]. SDK allows the developer to quickly develops the applications with ease [6]. Creating a SDK for remote collaboration between AR and VR with mobile games has several advantages: (1) leverage the advantages of both AR and VR together, (2) allow users to experience AR and VR in the mobile games with affordable price, and (3) able to develop AR and VR mobile games without programming knowledge with ease. Based on the current existing SDK has been produced for Unity, it does speed up the development process. The application such as mobile games are limited to a single platform such as AR alone and VR alone due to the hardware restriction and lack of collaborative plugin or modules. The collaboration of AR and VR technologies only focus on either AR or VR independently. Most AR or VR have been developed to only available with the particular and specific devices but very less in mobile devices. Therefore, this research aims to provide a SDK to enable a remote collaboration between AR and VR with mobile games.

2. Literature review

2.1. Augmented reality

Three main characteristics of AR [7] are: (1) combines both real and virtual entity, (2) interactive in real time, and (3) must be registered in 3-Dimensional (3D). AR superimposed virtual objects into the real world. Besides images, many other computer-generated information can be superimposed too [1]. There are several tracking techniques used in AR such as image feature detection, plane detection and object recognition. The development of AR in mobile devices continue to grow and received numbers of investments from top technology companies like Google, and Facebook. Some AR SDK that are compatible with the Unity3D game engine are Google ARCore, Vuforia, Wikitude, EasyAR, ARToolKit and ARKit. In this research, we implement the Vuforia SDK as it provides the natural feature tracking [8] that align with the tracking technique implemented in this research and it is compatible to works with mobile phone. Table 1 shows the comparison of AR SDKs.

Table 1. Comparison of AR SDKs.

	Vuforia	Wikitude	EasyAR	Kudan	ARToolKit	Maxst	Apple ARKit	XZIMG
Licence	Free, Commercial	Commercial	Free, Commercial	Free, Commercial	Free Open Source	Free, Commercial	Free	Free, Commercial
Supported platforms	Android, iOS, UWP	Android, iOS	Android, iOS, UWP, macOS	Android, iOS	Android, iOS, Linux, Windows, macOS	Android, iOS, Windows, macOS	iOS	Android, iOS, Windows
Smart glasses support	+	+	-	-	+	+	+	-
Unity support	+	+	+	+	+	+	+	+
Cloud recognition	+	+	+	-	-	-	+	-
3D recognition	+	+	+	+	-	+	+	-
Geolocation	+	+	-	-	+	-	+	-
SLAM	-	+	+	+	-	+	+	-

2.2. Virtual reality

There are four major key elements [9] that should be included in VR experience: (1) experience of being in the virtual world, (2) sense of feeling immersion, (3) receiving sensory feedback from the user input, and (4) interactivity. VR convince a human's brain about the existence of the computer-generated world. It combines the display technology with a 3D model in real-time providing an interactive environment [10]. There are several popular VR headsets, such as Google Cardboard, VR box, Oculus Go and HTC Vive. The built-in accelerometer in the mobile devices is often used to detect the head-position for generating movement or interact in the virtual world [11]. VR in the mobile device with headset such as Google Cardboard that is made with inexpensive materials and cable-less has gained its popularity among the consumers due to its affordable and portable characteristic [12]. Some VR SDK that are compatible with Unity3D game engine are Google Cardboard VR SDK, OpenVR SDK, SteamVR SDK and Oculus Mobile SDK. Table 2 shows the typical VR gears consist of mobile, standalone, and roomscale. Mobile and standalone have a boundary limits to 3 degree of freedom (DoF). While roomscale is enable 6 DoF. In this research, we use cardboard with high latency and low field of view (FoV) due to the gears is low in cost and its mobility to work with mobile device that comes with the built-in accelerometer that can be used to detect user head-position in the virtual environment [11].

Table 2. Typical VR gears.

2.3. Remote collaboration
The collaboration between the users can be achieved through: (1) face-to-face collaboration or (2) remote collaboration. All the VR users are required to wear a HMD that fully hide the vision of the real world's environment. Hence, the face-to-face collaboration is unachievable where the fusion of AR and VR can only be achieved through remote collaboration [13]. The remote collaboration between AR and VR refers to a mixed platform collaboration that enables multi-users with a different interface, either AR or VR interfaces to interact together when the users are located at different location [14]. The major goals of the collaboration in AR and VR are to change the human's sense allowing users to feel co-present through immersion of 3D environment while collaborating with each other [3]. The remote collaboration system of VR and AR (CoVAR) is discussed in [14] and [15].

The networking system connects the users via the network in a collaborative interface. The most renowned networking systems in the Unity3D are: (1) Photon Unity Networking (PUN) and (2) Unity Network (UNET). However, UNET may induces a higher latency in transmitting networked data, increase the device's workload hence reduce the performance in compares to PUN which might cause registration and interaction error in the remote collaboration [18].

2.4. Software development kit (SDK)
The SDK contains components, tools or functions that can be used to build an entirely new program, system or application. A successful SDK can reduce the developing time and eases in the creation process [19]. Each SDK has its own license type, targeted platforms, programming languages and related features and functionalities [6].

3. Methodology
After the literature review has been completed, the waterfall model is used to actualize this work, the following subsections explained about the methodology.

3.1. Analyze remote collaboration between AR and VR
The first phase focuses on the preliminary analysis of the remote collaboration in AR and VR. The analyses are conducted to have a better understanding of the whole methodology. The comparison

between the related tools has been made to ease in choosing the suitable tool during the prototyping in the later phase. Some existing projects have been studied to gain a better understanding of this work and the experiments. Several suitable options such as AR SDK and tracking technique, VR SDK and the headset, the networking service and the development platform are discussed. Besides, the gameplay for both card matching game and chess game are also explored to ease in scripting the game logic.

3.2. Develop the remote collaborative AR and VR mobile games

Based on the result analyzed in Phase 1, the suitable AR SDK, VR SDK and collaborative technology are determined in this phase. The Vuforia AR SDK is used to developed the AR interface. The image feature detection provided in the Vuforia is used for rendering the game in AR. Camera in mobile device constantly detects the marker and the 3D virtual objects are rendered and superimposed onto the real world accordingly. Figure 1 shows the AR tracking process.

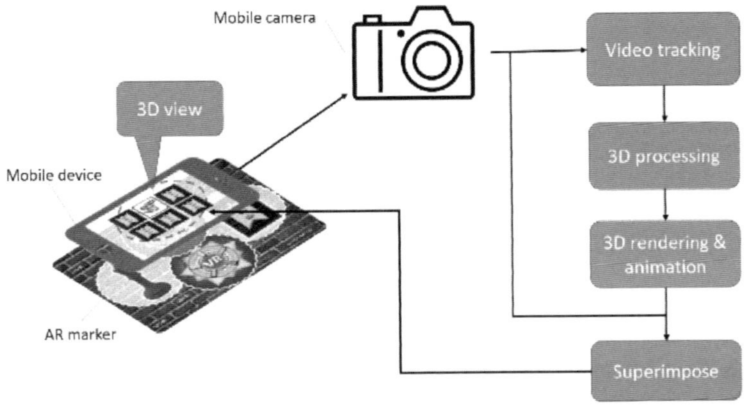

Figure 1. AR tracking process.

Google Cardboard VR SDK is used to developed the VR interface. The mobile device with gyroscope traces the x, y and z rotation and angular velocity. The VR system then match the in-game camera to follow the mobile's gyroscope and ray selection is used to select virtual objects in the virtual environment.

A mobile device is placed in the VR box to experience the 3D view through the mobile device's screen. Figure 2 shows the VR tracking process.

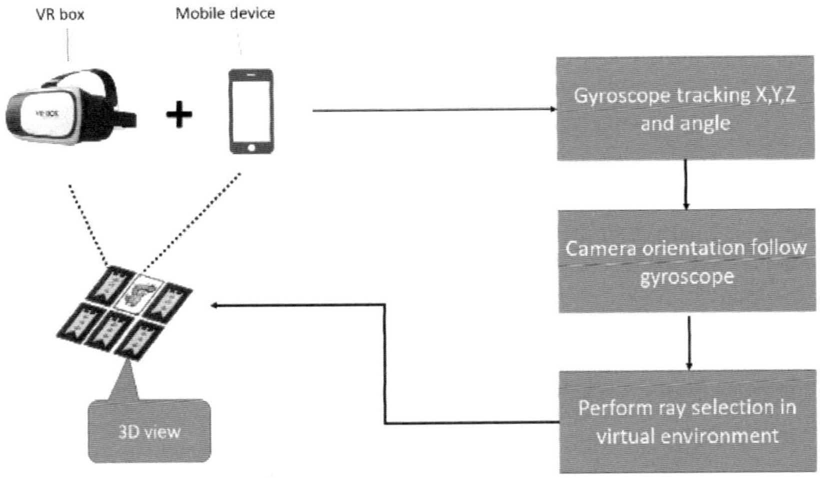

Figure 2. VR tracking process.

PUN is chosen to be the collaborative interface to connect both AR and VR users remotely. The algorithm for the collaboration between AR and VR are scripted such that the users from both AR and VR interface are able to take turns to interact with the same virtual object in real-time.

Figure 3 shows the system architecture. The whole methodology implementation is conducted as illustrated in the system architecture to actualized the remote collaboration between AR and VR.

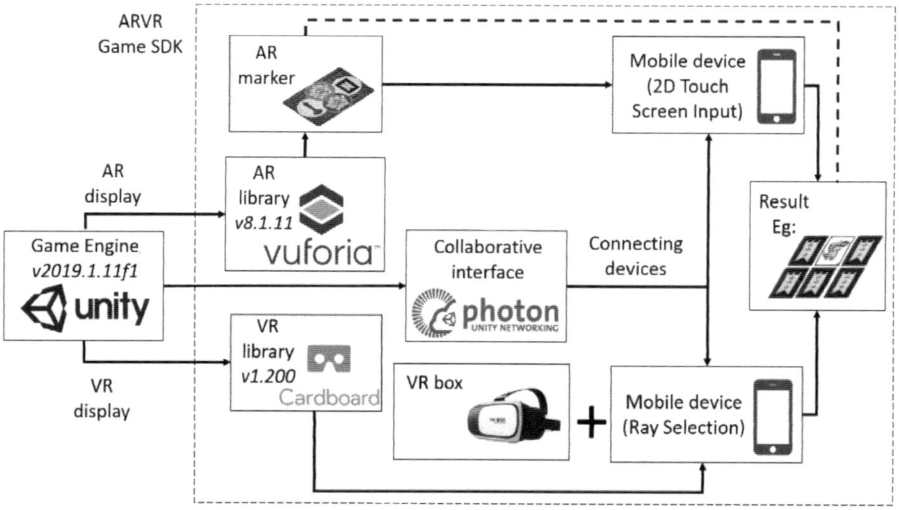

Figure 3. System architecture.

3.3. Construct the Mobile Game SDK with remote collaboration between AR and VR

The mobile games with remote collaboration between AR and VR that have been developed in the previous phase is constructed as SDK to achieve the research aim. Figure 4 shows the flow chart of the SDK.

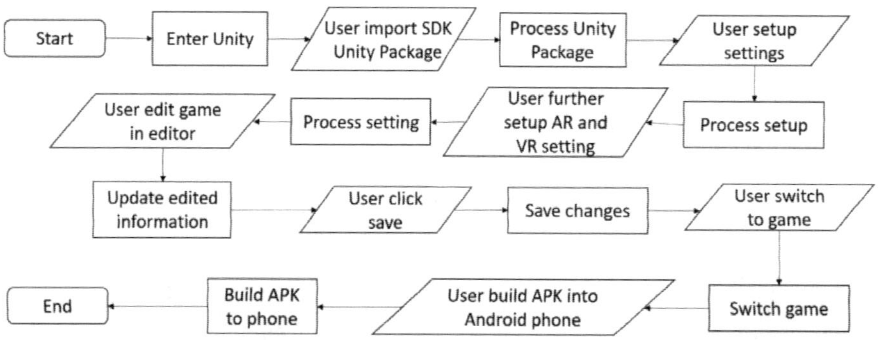

Figure 4. SDK flow chart.

The SDK is constructed such that the users can easily build their own collaborative AR and VR games into mobiles and play immediately without writing a single line of code. Besides, the users are allowed to change the desired texture or colour of the game object with ease

4. Implementation

The implementation of remote collaboration between augmented and virtual reality in mobile games is carried out according to the design as shown in the system architecture.

4.1. Develop card matching game

The card play is designed according to the traditional card game's concept and rules. There are a total of 3 pairs of cards facing down, that make it a total of 6 cards. A CardSpawner class acts as a controller to spawn all the 6 cards to the respective position. A GameController class acts as the card matching game controller. This class controls the game algorithms like determining cards' flipped, determining card matched and handling users' turns.

Figure 5 shows the gameplay flow chart for card matching game. Once the game start, the algorithm is scripted to spawn the card and set the user turn. The user then opens the first card and continues by opening the second card. If both cards opened by the user is the same, user gets the point and both cards are removed from the game. If both cads opened by the user happened to be different, both cards turn back faced down and it switches to the other user's turns. After the cards being removed, if there is no card left on the game, the game ends with the user with most points win. However, if there are cards left on the game, it continues by switching to the other user's turns.

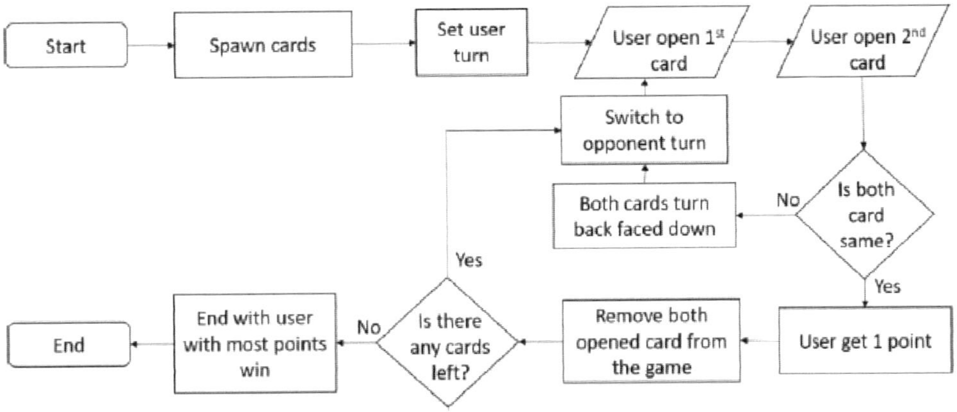

Figure 5. Card Matching game flow chart.

4.2. Fusion between AR and VR

PUN connects the users between AR and VR games. When running the application, a connection is sent to the Photon Server. A room is allocated in Photon Cloud for Master Client, the user who request to create a new room. The user that join the room later is known as the Client and is connected to the Photon Cloud server after joining the room. A real time synchronization of the gameplay's status is implemented to both users across the network by applying the Photon View component. Remote Procedure Calls (RPC) synchronized the function to be executed so that when a user performs an action that called a function, that particular function on the other user's side is called as well. This allows both AR and VR interface to be synchronized across the network via internet connection.

Figure 6 shows the fusion between AR and VR users. Both AR user and VR user are required to connect to the server before joining the room. After the game starts, AR user needs to scan the AR marker with the mobile device, while the VR user is required to wear a VR box with a mobile device placed inside to play the game.

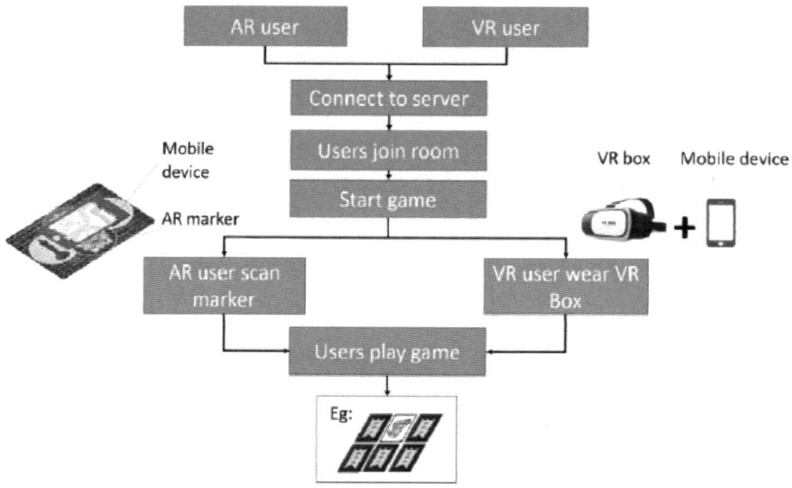

Figure 6. Fusion between AR and VR.

Figure 7 shows the collaboration between two users, the AR user and the VR user. The flow of the hardware and software works to enable the fusion of AR and VR is shown. In the hardware, VR box is attached with the mobile device to run VR application, while networking is required to connect VR user to the AR user that consists of a AR marker and a mobile device. For user interaction, gaze interaction is used for VR user and touch screen input is used for AR user.

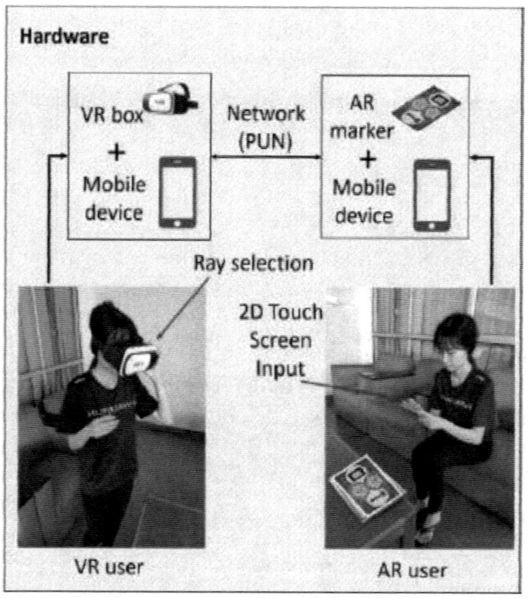

Figure 7. Collaborative between two users in AR and VR fusion.

4.3. User Interaction for AR and VR users

Multi-user interaction is defined as the action of more than one user able to interact concurrently in the environment. However, the basic multi-user interaction approach is incapable of distinguishing individual users, which leads to conflicts in collaborative tasks that regularly happen. Figure 8 illustrates the multi-user interaction workflow in the shared space context of integration between two different interface settings of AR space and VR space.

Multi-user interaction implements the client-server architecture to integrate AR and VR users in a shared space, where either the AR or VR user can be the host server. The shared space is independent of the system. Its position and rotation data are transmitted to one another between the AR or VR users via a network-enabled by internet protocol in real-time. Thus, the system eventually provides real-time 3D space interaction in VR and 2D touch input interaction for handheld AR applications. The integration phase is separated into two categories which are a server and a client. The server is a host computer that handles all connection requests from clients, maintains and manages the virtual content shared by users, and processes data storage and retrieval. The manipulated object properties are later updated through the network and are sent to the server where the shared space metaphor is situated. The virtual object properties consist of information that holds the transformation value of the virtual object and the boolean value that notify whether the object is currently manipulated by another user or not. The boolean value assists in organizing the interaction cue of a multi-user interaction in a collaborative system for the interaction to be smooth and sequential. The feedback received from the server as the new properties of the virtual object information is later updated and synchronizes on the client-side of the respective user spaces, either AR or VR.

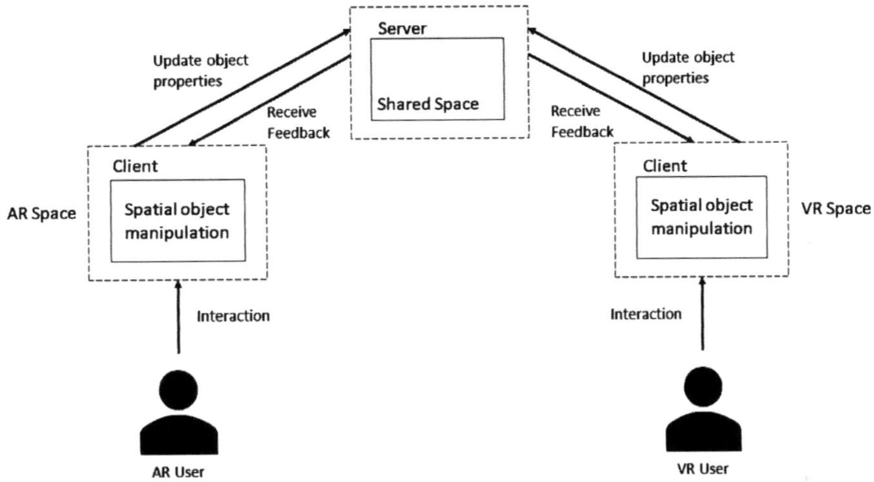

Figure. 8 Multi-user interaction workflow in the shared collaborative space.

Table 3. Interaction techniques for users in AR and VR fusion.

	AR	**VR**
Technique	Touchscreen interaction	Gaze interaction
Description	A touchscreen allows the user to interact by using their finger or stylus on a display device.	The gaze input uses a visual reticle to aim for user selection.
Input	Tap 	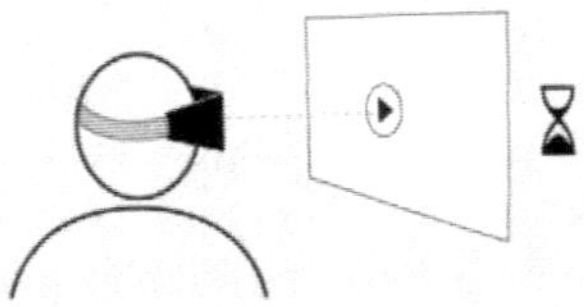

Table 3 shows the interaction techniques, when the user moves their head, a ray pointing was triggered the selectable object. The ray pointing was moving corresponding to user's head movement. *PointerEnter* function was called and the gaze was loading to fill up clockwise loading to complete the timer. The fill-up speed was according to the time set in the *ARVRInput* class. *PointerExit* was triggered when the loading was completed within the time.

4.3.1. Gaze Input

This study uses a gaze input, where the study does not have to use a magnetic trigger. Magnetic trigger uses a magnetometer that built-in in the smartphones and also used for a compass. Gaze input is when the user just keeps looking for a certain time (in our tests, 1.2 seconds (+ 0.2 seconds for final decision) worked best) on a button and then it gets selected, as shown in Fig 9.

Figure 9 (a) shows the targeted object is not yet hit by the gaze input, Figure 9 (b) shows the user is able accurately target the objects by moving their gaze. This interaction uses a visual reticle where there is a ray pointing used as a visual aid to track the object. Displaying a reticle, user can do fine targeting tasks and can help them to aim. This method displays the reticle only when the user approaches a target that they can activate as presented in Figure 9 (b). It is obvious hover states when the user moves their head to aim the targeted object.

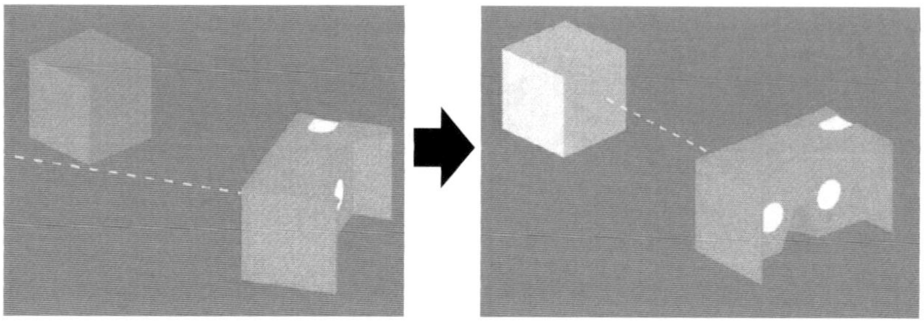

(a) The gaze does not look at the object (b) The gaze has triggered the object

Figure 9. Gaze input for user interaction in VR.

4.3.2. Touch Input

The mobile technologies nowadays typically allow a user to interact with a screen. Some technologies may work with only the user's finger, and other technologies may allow other tools, like a stylus. The mobile technologies have a capacitive touch screen which a screen that coated with a special material. When user touches a capacitive touch screen, there was a point of contact to indicate where the fingers have touched the screen.

In this study, we use touchscreen interaction since it was common interaction used for user to interact with AR content. Tapping is a single touch or briefly tap on the screen surface with a finger. When compared to an old-fashioned computer, a tap is the same as clicking with a mouse.

5. Results

The result from the experiment is a working SDK that acts as a plugin to Unity. This SDK can be used to develop mobile games where the remote collaboration of AR and VR were involved in the game. The Android mobile games were the card matching game and the chess game. These games support multiplayer where the users of the game were a AR user and a VR user.

During the gameplay, the User Interface (UI) for both AR and VR users were different. For an AR user, the game objects only appeared when the marker was detected. For card matching game, all the six cards were laid facing down during the game start, while chess game has all chessman arranged accordingly on the chessboard. During the gameplay, AR user interacts with the game using touchscreen.

The SDK developed adds a new menu item named "*ARVRGame*" to the main menu in the Unity3D game engine as shown in Figure 10 (a), and Figure 10 (b) shows the materials that available. Figure 10 (c) shows the matching game editor when the user started to build the card matching game.

IOP Conf. Series: Materials Science and Engineering **979** (2020) 012003 doi:10.1088/1757-899X/979/1/012003

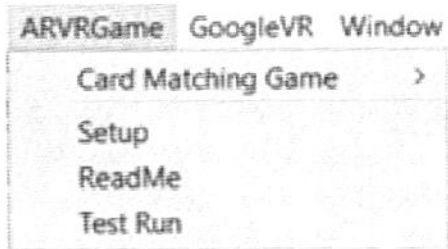

(a) SDKs in Unity IDE

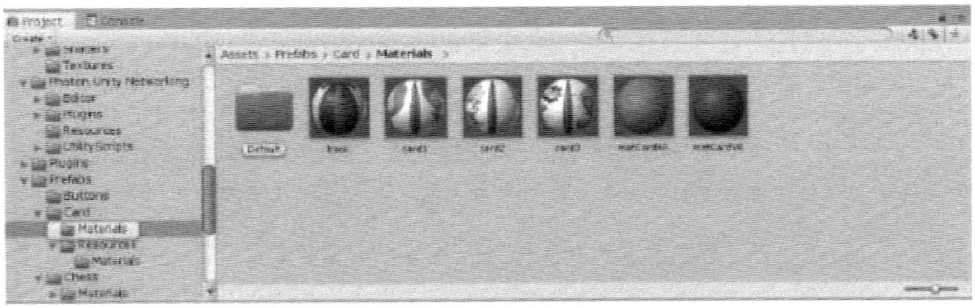

(b) Materials for Matching Card

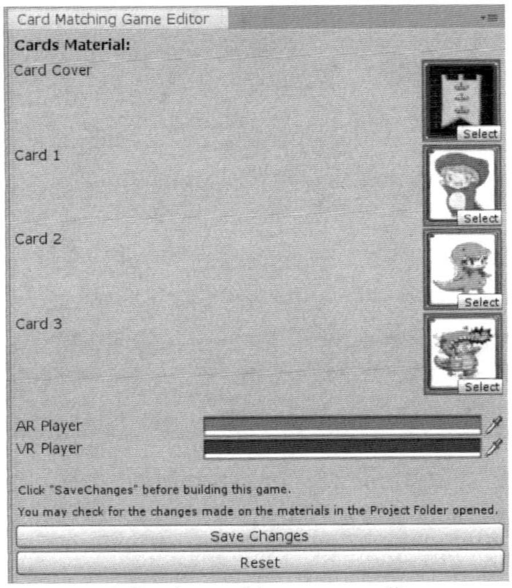

(c) Card matching game editor

Figure 10. Game SDK with remote collaboration AR and VR.

Figure 11 shows the AR matching game, and AR user's in-game decided to quit. The confirmation panel appears to ask the user whether want to choose cancel or quit. Figure 12 (a) shows the VR user's in-game UI for card matching game and Figure 11 (b) shows the VR user's in-game UI in card matching game when the quit button was being pressed.

IOP Conf. Series: Materials Science and Engineering **979** (2020) 012003 doi:10.1088/1757-899X/979/1/012003

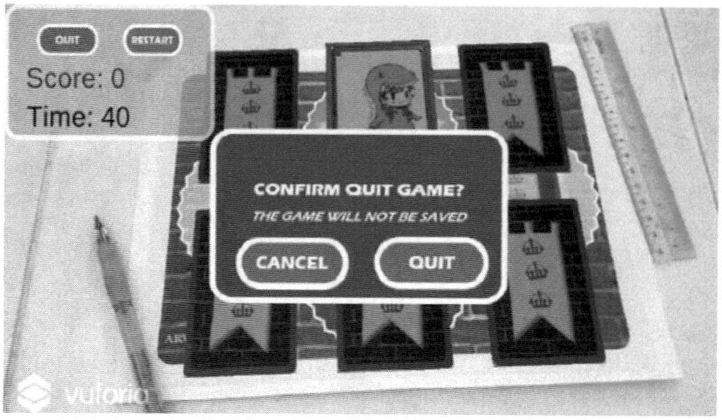

Figure 11. Card matching game in AR.

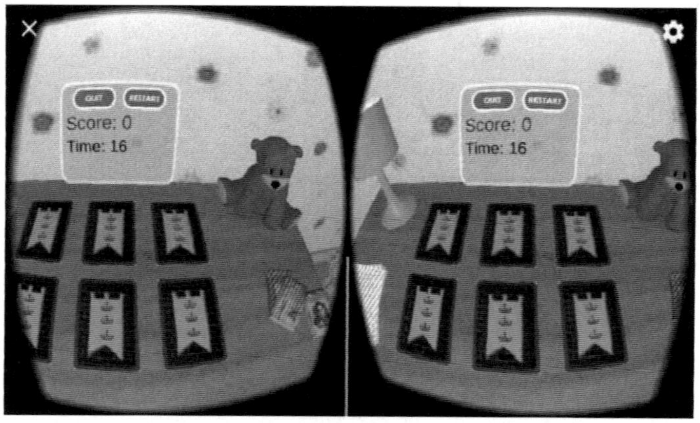

(a) Matching card game in VR using Cardboard headset

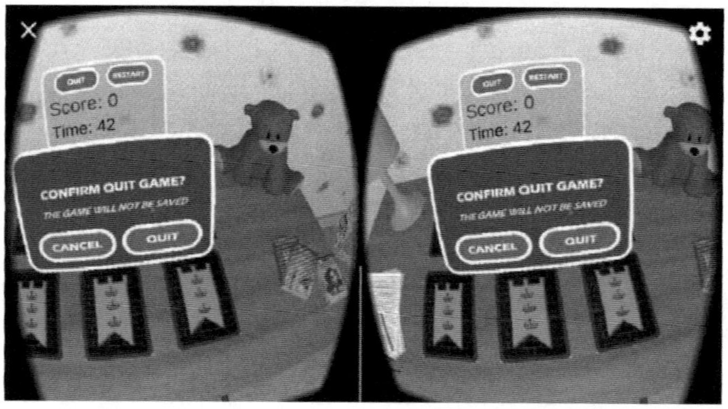

(b) User chooses to quit; the confirmation panel appears.

Figure 12. Card matching in VR.

Acknowledgement

We would to express our appreciation to Mixed and Virtual Reality Laboratory (mivielab), ViCubelab at Universiti Teknologi Malaysia (UTM) for the facilities and technical supports. Special thanks to UTMHIR Grants (Q.J130000.2451.04G77) for the funding.

References

[1] M. N. A. Nor'a and A. W. Ismail, "Integrating Virtual Reality and Augmented Reality in a Collaborative User Interface," *Int. J. Innov. Comput.*, vol. 9, no. 2, pp. 59–64, 2019
[2] A. A. Oyelude, "Virtual reality (VR) and augmented reality (AR) in libraries and museums," 2018
[3] T. Piumsomboon, G. A. Lee, and M. Billinghurst, "Snow Dome : A Multi-Scale Interaction in Mixed Reality Remote Collaboration," no. May, 2018
[4] L. Muñoz-Saavedra, L. Miró-Amarante, and M. Domínguez-Morales, "Augmented and virtual reality evolution and future tendency," *Appl. Sci.*, vol. 10, no. 1, 2020
[5] S. Benbelkacem, N. Zenati-Henda, D. Aouam, Y. Izountar, and S. Otmane, "MVC-3DC: Software architecture model for designing collaborative augmented reality and virtual reality systems," *J. King Saud Univ. - Comput. Inf. Sci.*, 2019
[6] A. Hanafi, L. Elaachak, and M. Bouhorma, "A comparative study of augmented reality SDKs to develop an educational application in chemical field," *ACM Int. Conf. Proceeding Ser.*, vol. Part F1481, 2019
[7] R. T. Azuma, "A Survey of Augmented Reality," vol. 6, no. 4, pp. 355–385, 1997
[8] Ulfah, S., Ramdania, D. R., Fatoni, U., Mukhtar, K., Tajiri, H., & Sarbini, A. (2020, June). Augmented reality using Natural Feature Tracking (NFT) method for learning media of makharijul huruf. In *IOP Conference Series: Materials Science and Engineering* (Vol. 874, No. 1, p. 012019). IOP Publishing.
[9] W. R. Sherman and A. B. Craig, "Introduction to Virtual Reality," *Underst. Virtual Real.*, pp. 5–37, 2003
[10] P. Vichitvejpaisal and T. Chotined, "Virtual Reality Photo Hunt Game Application for Amblyopia Patient," *ICSEC 2017 - 21st Int. Comput. Sci. Eng. Conf. 2017, Proceeding*, vol. 6, pp. 57–60, 2018
[11] W. Hürst and M. Helder, "Mobile 3D graphics and virtual reality interaction," p. 1, 2011
[12] S. Yoo and C. Parker, "Controller-less Interaction Methods for Google Cardboard," no. August, pp. 127–127, 2015
[13] S. Ranade, M. Zhang, M. Al-sada, J. Urbani, and T. Nakajima, "Clash Tanks : An Investigation of Virtual and Augmented Reality Gaming Experience," pp. 0–5, 2017
[14] T. Piumsomboon, A. Dey, B. Ens, G. Lee, and M. Billinghurst, "CoVAR: Mixed-Platform Remote Collaborative Augmented and Virtual Realities System with Shared Collaboration Cues," in *Adjunct Proceedings of the 2017 IEEE International Symposium on Mixed and Augmented Reality, ISMAR-Adjunct 2017*, 2017, pp. 218–219
[15] T. Piumsomboon *et al.*, "Mini - Me : An Adaptive Avatar for Mixed Reality Remote Collaboration," *Proc. 2018 CHI Conf. Hum. Factors Comput. Syst. - CHI '18*, pp. 1–13, 2018
[16] A. A. Kubena, "Pexeso ('Concentration game') as an arbiter of bounded-rationality models," *28th Int. Conf. Math. Methods Econ. 2010, Pts I Ii*, pp. 377–380, 2010
[17] M. Diaz, M. Alencastre-Miranda, L. Munoz-Gomez, and I. Rudomin, "Multi-user Networked Interactive Augmented Reality Card Game," pp. 13–16, 2006
[18] C. S. Yusof, T. S. Low, A. W. Ismail, and M. S. Sunar, "Collaborative augmented reality for chess game in handheld devices," *2019 IEEE Conf. Graph. Media, GAME 2019*, pp. 32–37, 2019
[19] A. Kolsanov, A. Nazaryan, A. Ivaschenko, and A. Kuzmin, "Intelligent SDK for 3D surgery simulation," *Proc. 2016 6th Int. Conf. - Cloud Syst. Big Data Eng. Conflu. 2016*, pp. 384–387, 2016
[20] D. Mardanbegi and T. Pfeiffer, "EyeMRTk: A toolkit for developing eye gaze interactive applications in virtual and augmented reality," *Eye Track. Res. Appl. Symp.*, 2019.
[21] R. Van Herk, J. Verhaegh, and W. Fontijn, "Espranto SDK: An adaptive programming environment for tangible applications," *Conf. Hum. Factors Comput. Syst. - Proc.*, pp. 849–858, 2009
[22] D. Amin, & S. Govilkar, (2015). Comparative study of augmented reality SDKs. International *Journal on Computational Science & Applications*, 5(1), 11-26.

IOP Conf. Series: Materials Science and Engineering **979** (2020) 012004 doi:10.1088/1757-899X/979/1/012004

Object selection and scaling using multimodal interaction in mixed reality

M Y F Aladin [1,2] *, A W Ismail [1,2], N A Ismail [1,2] and M S M Rahim [1,2]

[1] Mixed and Virtual Environment Research Lab (mivielab), ViCubeLab, Universiti Teknologi Malaysia, 81310 Johore, Malaysia
[2] School of Computing, Faculty of Engineering, Universiti Teknologi Malaysia, 81310 Johor, Malaysia

* yahyafekri@gmail.com

Abstract. Mixed Reality (MR) is the next evolution of human interacting with the computer as MR has the ability to combine the physical environment and digital environment and making them coexist with each other. Interaction is still a huge research area in Augmented Reality (AR) but very less in MR, this is due to current advanced MR display techniques still not robust and intuitive enough to let the user to naturally interact with 3D content. New techniques on user interaction have been widely studied, the advanced technique in interaction when the system able to invoke more than one input modalities. Multimodal interaction undertakes to deliver intuitive multiple objects manipulation with gestures. This paper discusses the multimodal interaction technique using gesture and speech which the proposed experimental setup to implement multimodal in the MR interface. The real hand gesture is combined with speech inputs in MR to perform spatial object manipulations. The paper explains the implementation stage that involves interaction using gesture and speech inputs to enhance user experience in MR workspace. After acquiring gesture input and speech commands, spatial manipulation for selection and scaling using multimodal interaction has been invoked, and this paper ends with a discussion.

1. Introduction

Mixed Reality (MR) is a technology where digital world coexists with our physical world. Digital information has the ability to understand our physical world making a virtual 3D robot can hide behind a physical couch and play hide and seek with us. Differs from Augmented Reality (AR) where virtual content can only be overlaid onto the physical world through a video stream. MR term originates from [1]. Since then, MR has grown so much more than just for display but covering environment understanding and interaction metaphor. MR, no longer fiction or computer-generated imaginary is made possible by the advancement of graphic processer, central processor, computer vision and input systems. To make sense the MR technology, a natural and real interaction needs to be implemented to enhance the user experience when interacting with the virtual content [2]. There are many natural input systems based on the human form, such as gaze, facial expressions, gesture, and speech. These natural inputs are crucial to let the user interact with virtual contents in the most natural manner. This paper explains about implementing multimodal interaction for MR using speech and gesture for MR interface. User interaction has been produced in AR, very less in the MR environment. Due to the advanced technique, multimodal inter-action has been recommended by most of the researchers due to the complementary inputs, and more than one input more user to interact. The remaining issues in user interaction, the user interaction needs to as natural as possible as agreed by [3], and the 3D object manipulation methods need to be as intuitive as possible [4]. An intuitive multimodal interaction technique in the MR environment can be achieved when the occlusion issue can be improved in this research. There are depth mapping errors during the capturing process, and the errors need to be minimized to correspond with the real object such as a user's hand.

2. Related Works

MRTouch [5] is a multitouch input solution for MR, and in their system, the user uses their physical environment as if they were touchscreens as shown in Figure 1. User is able to use their finger to interact with the virtual inter-faces affixed to surfaces. He incorporates Microsoft HoloLens as dis-play device and takes advantage of HoloLens gesture recognition mod-ule to produce their multitouch interaction.

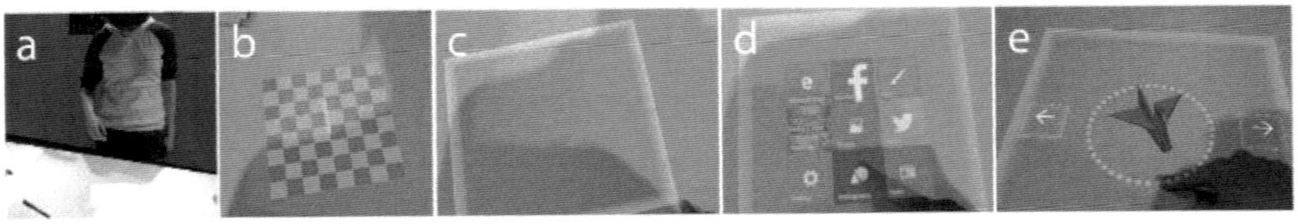

Figure 1. MyTouch application interfaces [5].

Another research by Meegahapola [6] produced a system that enhanced in-store shopping experience through a smart phone or in short SPMRA – phone-based MR application. This application, instead of using a costly device such as HoloLens, incorporate a mobile phone as the headset by using a Samsung Gear VR or Google Cardboard. This application claims to integrate multimodal interaction using gaze and physical button on the side of the Samsung Gear VR. CETA [7], MR system for school-aged children to enhance mathematical learning using Tangible User Interaction (TUI). The system uses a low-cost android tablet, a mirror, a holder to hold the tablet, and a set of rectangle wooden blocks that acts as the tangible interface. While in architecture design, [8] has designed a prototype for multimodal to generate displays in 3D environments so users can use a variety of devices to interact in the same networked environment. However, it was fully immersive VR environment not in MR. Same as [9], they proposed a multimodal interaction to enhance creativity during designing tasks with a pulsing microphone icon to give participants feedback and indicate when they can interact in a VR environment.

In the AR environment, Ismail et al. [10] have produced a system called Multimodal Interaction in Augmented Reality or MIAR for short. The input modalities used in MIAR is natural hand gesture and speech. A leap motion device is used to track and recognize the gestures performed by the user and a microphone is used to capture the user's speech command. Piumsomboon et al. [11] adopted the use of gesture and speech input modalities in a system called Gesture-Speech Interface for Augmented Reality or G – SIAR for short and the project setup is as illustrated in Figure 2. The free-hand gesture interaction techniques or Grasp-Shell and multimodal interaction technique Gesture-Speech is constructed to be compared and evaluated.

Figure 2. G-SIAR setup.

Table 1 shows the related works in multimodal that has been listing by year in 2015 until 2020. The five years of works indicate that the projector-based started with mobile-based and headed to see-through headset and projectors. Table 2 shows the gesture interaction that commonly invoked in MR/AR. There are two categories, gesture input using real hand and fingertip using marker-based. Marker-based is when the user's hands were attached with the black and white physical marker and interactions such as pointing, grabbing, push and pressing can be defined.

Table 1. Previous works on multimodal interaction in AR and MR from 2015-2020.

Year	Research	Interaction technique	Display
2015	Vision-Based Technique and Issues for Multimodal Interaction in AR [12]	Face, Gesture, Speech	See-through based, Mobile-based
	MIAR: Gesture and Speech Input in AR Environment [10]	Speech, Gesture	PC-based
	ARZombie [13]	Face, Gesture	Mobile-based
2016	In Situ CAD Capture [14]	Speech, Gesture	Mobile-based
	Semantics-based Software [15]	Speech, Gesture	Mobile-based
2017	Adaptive Multimodal Input [16]	Speech, Motion, Gesture, Adaptive	Mobile-based
	Multimodal Interaction in Augmented Reality [17]	Speech, Gesture	Mobile-based
	Mobile AR with the Hololens [18]	Speech, Gesture, Tangible	See-through based
2018	HandsInTouch [19]	Gesture, Sketch	See-through based, Pc-based
	Space Tentacles [20]	Gesture, Speech	Pc-based
	A Multimodal System involves AR, Gestures, and Tactile Feedback [21]	Gesture, Tactile	See-through based
2019	Multimodal Driver Interaction [22]	Gesture, Gaze, and Speech	See-through based
	Multimodal Referring Expressions with Mixed Reality [23]	Gesture, Speech	See-through based
	A multimodal interface for virtual information environments [24]	Eye gaze, Gesture, Speech	Pc-based
2020	Bio-Holograms in Mixed Reality [25]	Gaze, Eye, Haptic	Projectors, HMD
	A 3D model of student interest during learning using multimodal fusion [26]	Face, Gaze	Pc-based, HMD

Table 2. Gesture Inputs and Techniques.

	Pointing	Grabbing	Push	Pressing
Gesture Input using real hand [27]	✓ A pointing gesture is recognized when the user touches their finger to the ground	✓ Grabbing is used to move objects by grabbing them and releasing them again	✓ Push object to change the position by pushing it in a direction	✗ Pressing gestures are a modification of pointing gestures and are used to interact with a virtual button
Fingertip using marker-based [28]	✓ A pointing gesture is recognized when the user touches their finger to the ground	✓ Grabbing is used to move objects by grabbing them and releasing them again	✗ Push object to change the position by pushing it in a direction	✓ Pressing gestures are a modification of pointing gestures and are used to interact with a virtual button

3. Implementation

There are three phases that have been carried out to design the MR interface that covers the areas as described in the following subsections.

3.1. Configuration and User Setting

The interaction for MR-Deco includes rotation, translation, and also texture manipulation, where the user can change the texture of the object. HMD is essential in this development. There are several interaction controllers can be used with the HMD such as Oculus Touch, Oculus Remote and also Joystick. As illustrated in Figure 3 (a), the workspace to set up VR display technology, there are listed the computer desktop, Oculus Sensor, Oculus Rift and Oculus Touch. Oculus Rift is chosen as the display device for users to experience and interact with the MR environment. Leap Motion controller is used to capture the user's hands. Leap Motion is a controller to provide a workspace with appropriate gesture inputs; it was attached to HMD (as in Figure 3 (b)). HMD was attached with Zed Mini RGB depth (RGB-D) camera. A Zed Mini depth camera for images does not use infrared; it instead uses two RGB-D cameras.

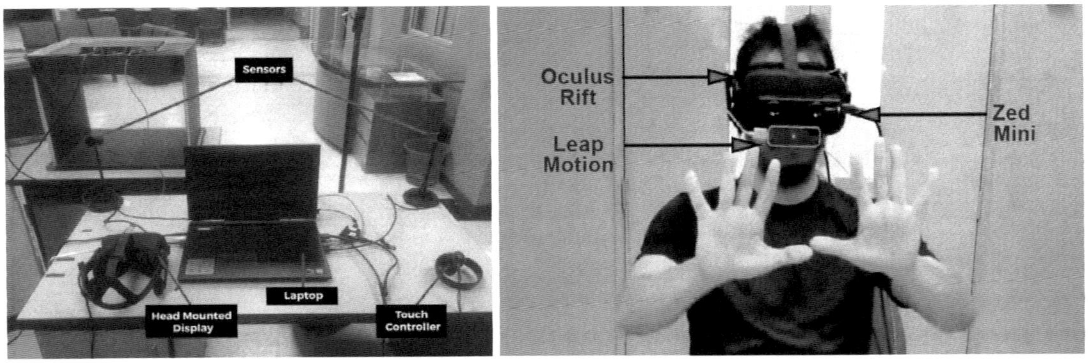

(a) Hardware for workspace setting (b) Leap Motion Controller to capture real hand

Figure 3. MR workspace setting.

3.2. Acquiring Gesture Inputs using Leap Motion

This study adopts the Leap Motion device to take advantage of the powerful hand tracking. The device works by having two cameras and three infrared LEDs that track infrared light that is outside the visible light spectrum. The sensor reads the data and later streamed via USB to Leap Motion library. Then the calculation takes place in the software to compute the finger positions and orientation. Then the prototype will take advantage of this information to be used as a gesture interaction. The interaction features in the prototype support two kinds of gesture – hover and grasp:

- With hover, the user can touch the 3D content, and a visual cue was projected to the user to show that the user's hands are occluded with the 3D content.
- With grasp, users are able to pick the 3D object and manipulate it. Users are able to translate and rotate the 3D object seamlessly and naturally as if the user holds a real physical object.

When the user moves the hands, the Leap Motion will track the movement in real-time. When hands are in specific space and range, the user is hovering their hands on the Leap Motion controller, and a recognition process will give the system access to its position with the orientation. As the hand inside the real-time gesture frame, the motion data is provided with a direction vector to its gesture input definitions. A gesture inputs may have numerous components such as pointing, pinch, grab, and stretch by using both hands. The pinch gesture input, for example, will imply the natural movements several times and other gestures, approximating stretch or shrink is in different axis and rotation, which can either use a single hand or both.

(a) Real environment (b) Palm up in MR (c) Palm down in MR

Figure 4. Both hand recognition in MR space.

Figure 4 shows the initial result to produce natural user interaction gesture interface output of the application system when user see-through the HMD. A virtual hand will be rendered in real-time according to the finger's positions and orientations. The user and his hands appear visible in MR space when gesture performs the action as shown in Figure 4, and the user was demonstrating the virtual hand palm up or palm down. Based on Figure 4 (d), the gesture is tracked by sensor-based tracking technique where hand gesture natural features are being tracked by the gesture tracking devices. Figure 4 (a) illustrates the user in the real-world environment, performing palm up gesture (as in Figure 4 (b)). With the palm up and palm down (as in Figure 4 (c)), the user can perform gesture interaction in the MR world.

3.3. Hand Gesture in MR
Figure 4 demonstrates with depth occlusion-free enabled; the virtual car still appears to occlude the real hand. Without occlusion-free, the virtual car is occluded by the user's hand. User to place their hand in front of the virtual object and user can still see their real hand, with-out being occluded by the virtual object. With depth occlusion-free enabled, the virtual pixels can be covered up by the real pixels, making placing a virtual car behind a real hand, would make the virtual car in-visible or disappear. Figure 5 (a) the real hand covers the virtual car; it still appears on occluded by it. In Figure 5 (b), the user's hand is behind the virtual car and still appear occluded by it. In this result, it enables the user to place their hand in front of the virtual object and user can still see their real hand, without being occluded by the virtual object.

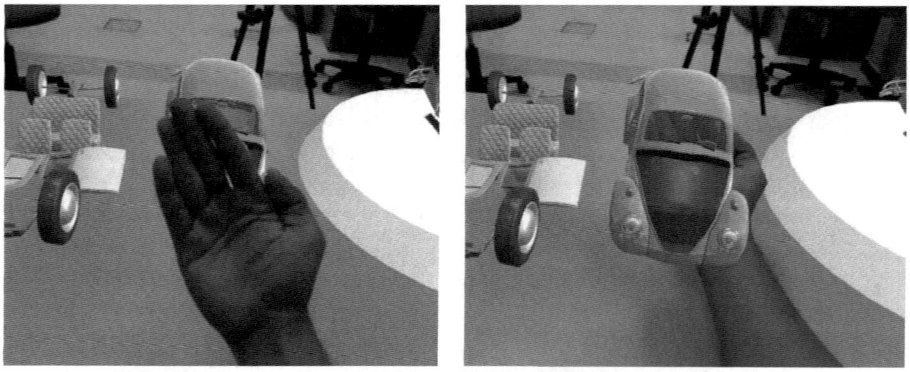

(a) Real hand covers the virtual object (b) Virtual object covers the real hand.

Figure 5. Occlusion issue in MR for real hand gesture.

4. Application

4.1. Selection Method

The Microsoft Speech API is used to synthesize the speech input, and the process will prompt the user's input from the window application. The speech synthesizer will send the input to be recognized by the application. The process ends with producing speech output. Further research is still continuing to indicate the efficient speech output that potential to complement the gestures input for multimodal interaction. The research also concerns the cues or user's turn in order to respond to the inputs provided by more than one input modality, gesture and speech.

The research also concerns the cues or user's turn in order to respond to the inputs provided by more than one input modality, gesture, and speech. For the proposed multimodal interaction, the user needs to engage with the 3D object using gesture first and then followed with speech command. Giving the speech command first and then invoking gesture input will not trigger the multimodal interaction. Thus, Figure 6 illustrates a flowchart for multimodal interaction technique can be performed in MR.

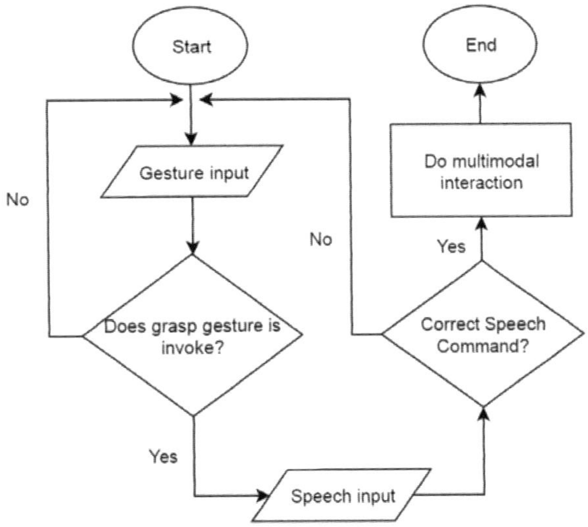

Figure 6. Flowchart for multimodal interaction technique.

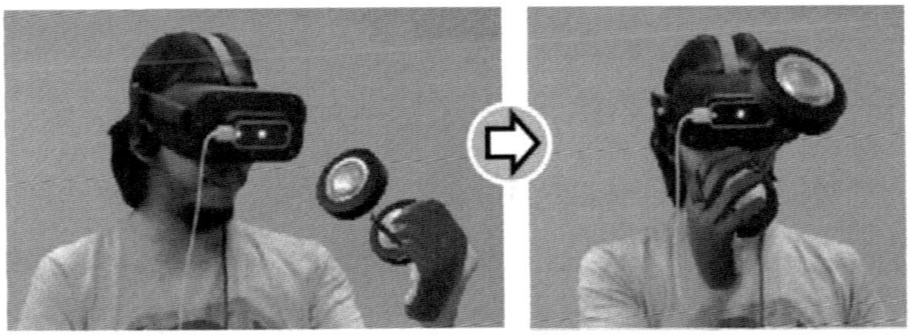

(a) (b)

Figure 7. Multimodal interaction using gesture and speech.

Figure 7 shows the representation for multimodal interaction using gesture and speech to be executed. Firstly, the system is checked for the presence of grasp or pinch gesture, and then next it was indicating that the user is currently holding the 3D object (as in Figure 7 (a)). If such gesture input is not existent, the system continues searching for the presence of those gesture input. Until the gesture is found, only then the system allows for speech input to activate. As in Figure 7 (b)), the user performs speech input

IOP Conf. Series: Materials Science and Engineering **979** (2020) 012004 doi:10.1088/1757-899X/979/1/012004

to increase scaling factor to two times and it can be seen the object size has been increased up to be bigger. In order to perform the task, the user needs to grasp the desired car parts, and user needs to speak out the right command ("increase size by one") in order for the prototype to recognize the command. If the speech command feeds the system with the correct speech command, multimodal interaction is performed. Otherwise, the system reverts the process back to gesture checking and searching. The significant reason why the cue needs to behave like this, if the gesture is able to work without speech, it will consider gesture-only interaction, and it won't be considered as a multimodal input.

In initial result to test a fascinating and interactive virtual space that lets users virtually explore the assembly of a car. This transformative technology allows users to grasp and manipulate virtual con-tents with simple hand movements that are easy to learn. The objects tested are several parts of the car, which are windows, tires, seat and body. Based on Figure 8, the user explores the VR car model using point gaze through gesture and using speech to invoke a task to change the parts of the car, which indicates the multimodal has been used. Figure 8 (a) shows the user selecting the body of the toy car, and while the car part is selected, the user utters the speech command "change this" as in Figure 8 (a) to perform MMI to change the car part: from the roof to roofless as seen in Figure 8 (b).

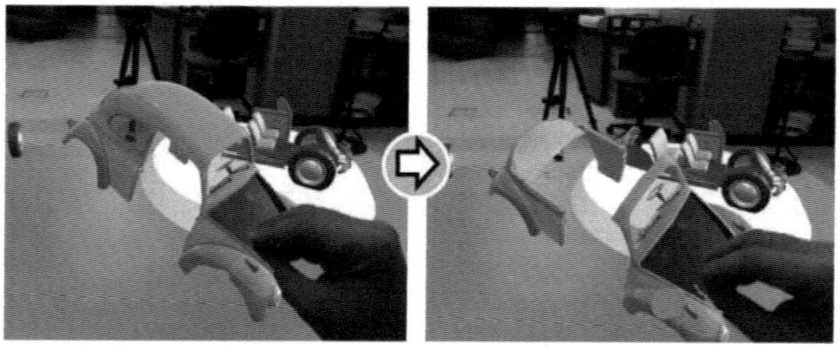

(a) (b)

Figure 8. User perform selection using multimodal interaction in MR.

4.2. Scaling

Similar to changing the car part, 3D object scaling also performed while the user selecting the car part: tires as in Figure 9 (a), and the user utters the speech command "increase the size by one" to scale the tires as illustrates in Figure 9 (b). The speech command can be uttered more than once if the user wants to invoke the same interaction. In order to perform uniform during object manipulation, the grasping and pinching gestures are not required, but instead, the hover gesture is used for this particular interaction. The user can perform hover gestures anywhere in the scene, and while it is performed, speech command "select all" need to be uttered to perform 3D objects selection, and all 3D objects in the scene will be selected.

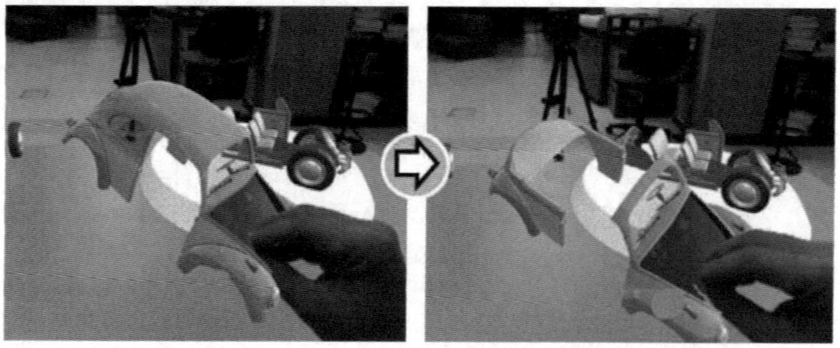

(a) (b)

Figure 9. User perform scaling using multimodal interaction in MR.

5. Conclusion and Future Works

This paper discusses the multimodal interaction technique using gesture and speech that the experimental setup to proposed multimodal in the MR interface. MR is rapidly becoming an enjoyable experience in exploring new techniques on tracking and interaction. The intuitive multiple objects manipulation with gestures was also involved in MR. The gesture data was captured by Leap Motion has been integrated to produce a robust hand tracking technique. The gesture is combined with speech inputs in MR to perform spatial object manipulations for MR.

During the gesture recognition process, the gesture inputs were captured by the Leap Motion device that is attached to the Oculus Rift. The head movement tracking corresponds to hand tracking. Speech input is acquired from the build-in microphone in Oculus Rift, and the speech recognition system is executed right after the enable button is pressed and will remain running until the disable button is pressed or the system is close. While for the speech input, the speech grammars list has been defined and combined with gesture interaction to perform multimodal, which has been analyzed to accept a maximum of four words. The grammar panel is required to give the user a handler and feedback to determine whether to enable or disable the speech recognition process. The speech commands need to store and load asynchronously.

Further extension for the future works, the use of current technologies such as Microsoft HoloLens, or Project North Star by Leap Motion [29] can improve the user experienced by the whole as this research only adopt an opaque Oculus Rift for display with the RGB-D camera attached to it. The user can directly see their surrounding without being occluded by an opaque device, and with this increased the field of view that the user sees from their surroundings. When working collaboratively, it is bound that sometimes different user is selecting the same 3D object, so in this case, there is a need for an algorithm to give priority to whoever touches the 3D object first. Therefore, the collaborative interface such as in [30] can be considered to implement multimodal interaction to leverage conventional interfaces.

Acknowledgement

We would like to express our appreciation to Mixed and Virtual Reality Laboratory (mivielab), ViCubeLab at Universiti Teknologi Malaysia (UTM) for the facilities and technical supports. We would like to convey our gratitude to Fundamental Research Grant Scheme (PY/2020/05052) for the funding.

References

[1] Ismail, A., & Noh, Z. (2008). Augmented Reality Theory and Applications. M. Shahrizal, & M. Najib, Advances in Computer Graphics and Virtual Environment, 88-105.

[2] Ismail A.W., Sunar M.S. Intuitiveness 3D objects interaction in augmented reality using the S-PI algorithm. TELKOMNIKA Indonesian Journal of Electrical Engineering. 2013. 11(7), pp.3561-3567.

[3] Ismail, A. W., & Sunar, M. S. (2015). Multimodal fusion: gesture and speech input in the augmented reality environment. In Computational Intelligence in Information Systems (pp. 245-254). Springer, Cham.

[4] Ismail, A. W. (2011). User Interaction Technique With 3D Object Manipulation in Augmented Reality Environment (Doctoral dissertation, Universiti Teknologi Malaysia).

[5] Xiao, R., Schwarz, J., Throm, N., Wilson, A. D., & Benko, H. (2018). MRTouch: adding touch input to head-mounted mixed reality. IEEE transactions on visualization and computer graphics, 24(4), 1653-1660.

[6] Meegahapola L., Perera I. Enhanced in-store shopping experience through smart phone-based mixed reality application. In 2017 Seventeenth International Conference on Advances in ICT for Emerging Regions (ICTer). IEEE. 2017. pp. 1-8.

[7] Marichal S., Rosales A., Perilli F.G., Pires A.C., Bakala E., Sansone G., Blat J. CETA: designing mixed-reality tangible interaction to enhance mathematical learning. In Proceedings of the 19th International Conference on Human-Computer Interaction with Mobile Devices and

Services. ACM. 2017. p. 29.
[8] Thiry, K. E., Wolloko, A., Kingsley, C., Flowers, A., Bird, L., & Jenkins, M. P. (2019, September). Designing Federated Architectures for Multimodal Interface Design and Human Computer Interaction in Virtual Environments. In International Conference on Human Systems Engineering and Design: Future Trends and Applications (pp. 404-410). Springer, Cham.
[9] Wolf, E., Klüber, S., Zimmerer, C., Lugrin, J. L., & Latoschik, M. E. (2019, October). " Paint that object yellow": Multimodal Interaction to Enhance Creativity During Design Tasks in VR. In 2019 International Conference on Multimodal Interaction (pp. 195-204).
[10] Ismail A.W., Sunar M.S. Multimodal fusion: gesture and speech input in the augmented reality environment. In Computational Intelligence in Information Systems. Springer, Cham. 2015. pp. 245-254.
[11] Piumsomboon T., Altimira D., Kim H., Clark A., Lee G., Billinghurst M. Grasp-Shell vs gesture-speech: A comparison of direct and indirect natural interaction techniques in augmented reality. In2014 IEEE International Symposium on Mixed and Augmented Reality (ISMAR). IEEE. 2014. pp. 73-82.
[12] Ismail, A. W., Billinghurst, M., & Sunar, M. S. (2015, August). Vision-based technique and issues for multimodal interaction in augmented reality. In Proceedings of the 8th International Symposium on Visual Information Communication and Interaction (pp. 75-82).
[13] Cordeiro, D., Correia, N., & Jesus, R. (2015, June). ARZombie: A mobile augmented reality game with multimodal interaction. In 2015 7th International Conference on Intelligent Technologies for Interactive Entertainment (INTETAIN) (pp. 22-31). IEEE.
[14] Sankar, A., & Seitz, S. M. (2016, September). In situ CAD capture. In Proceedings of the 18th International Conference on Human-Computer Interaction with Mobile Devices and Services (pp. 233-243).
[15] Fischbach, M., Wiebusch, D., & Latoschik, M. E. (2016, March). Semantics-based software techniques for maintainable multimodal input processing in real-time interactive systems. In 2016 IEEE 9th Workshop on Software Engineering and Architectures for Realtime Interactive Systems (SEARIS) (pp. 1-6). IEEE.
[16] Abidin, R. Z., Arshad, H., & Shukri, S. A. I. A. (2017). A framework of adaptive multimodal input for location-based augmented reality application. Journal of Telecommunication, Electronic and Computer Engineering (JTEC), 9(2-11), 97-103.
[17] Chen, Z., Li, J., Hua, Y., Shen, R., & Basu, A. (2017, October). Multimodal interaction in augmented reality. In 2017 IEEE International Conference on Systems, Man, and Cybernetics (SMC) (pp. 206-209). IEEE.
[18] Zimmer, C., Bertram, M., Büntig, F., Drochtert, D., & Geiger, C. (2017). Mobile Augmented Reality Illustrations that entertain and inform: Design and Implementation issues with the Hololens. In SIGGRAPH Asia 2017 Mobile Graphics & Interactive Applications (pp. 1-7).
[19] Huang, W., Billinghurst, M., Alem, L., & Kim, S. (2018, December). HandsInTouch: sharing gestures in remote collaboration. In Proceedings of the 30th Australian Conference on Computer-Human Interaction (pp. 396-400).
[20] Zimmerer, C., Fischbach, M., & Latoschik, M. E. (2018, March). Space Tentacles-Integrating Multimodal Input into a VR Adventure Game. In 2018 IEEE Conference on Virtual Reality and 3D User Interfaces (VR) (pp. 745-746). IEEE.
[21] Chan, W. P., Quintero, C., Pan, M., Sakr, M., Van der Loos, H. M., & Croft, E. (2018). A Multimodal System using Augmented Reality, Gestures, and Tactile Feedback for Robot Trajectory Programming and Execution. In ICRA Workshop on Robotics in Virtual Reality 2018.
[22] Aftab, A. R. (2019, October). Multimodal Driver Interaction with Gesture, Gaze and Speech. In 2019 International Conference on Multimodal Interaction (pp. 487-492). ACM.
[23] Sibirtseva, E., Ghadirzadeh, A., Leite, I., Björkman, M., & Kragic, D. (2019). Exploring

Temporal Dependencies in Multimodal Referring Expressions with Mixed Reality. arXiv preprint arXiv:1902.01117.

[24] Hansberger, J. T., Peng, C., Blakely, V., Meacham, S., Cao, L., & Diliberti, N. (2019, July). A multimodal interface for virtual information environments. In International Conference on Human-Computer Interaction (pp. 59-70). Springer, Cham.

[25] Romanus, T., Frish, S., Maksymenko, M., Frier, W., Corenthy, L., & Georgiou, O. (2020). Mid-Air Haptic Bio-Holograms in Mixed Reality. arXiv preprint arXiv:2001.01441.

[26] Luo, Z., Jingying, C., Guangshuai, W., & Mengyi, L. (2020). A three-dimensional model of student interest during learning using multimodal fusion with natural sensing technology. Interactive Learning Environments, 1-14.

[27] Buchmann, V., Violich, S., Billinghurst, M., & Cockburn, A. (2004, June). FingARtips: gesture based direct manipulation in Augmented Reality. In Proceedings of the 2nd international conference on Computer graphics and interactive techniques in Australasia and South East Asia (pp. 212-221).

[28] Dingler, T., Funk, M., & Alt, F. (2015, June). Interaction proxemics: Combining physical spaces for seamless gesture interaction. In Proceedings of the 4th International Symposium on Pervasive Displays (pp. 107-114).

[29] Motion, L. (2019). Project North Star is Now Open Source-Leap Motion Blog. Leap Motion Blog.

[30] Nor' a, M. N. A., & Ismail, A. W. (2019). Integrating Virtual Reality and Augmented Reality in a Collaborative User Interface. International Journal of Innovative Computing, 9(2).

IOP Conf. Series: Materials Science and Engineering **979** (2020) 012005 doi:10.1088/1757-899X/979/1/012005

3D telepresence for remote collaboration in extended reality (xR) application

F E Fadzli [1,2] *, M S Kamson[1,2], A W Ismail [1,2] and M Y F Aladin [1,2]

[1] Mixed and Virtual Environment Research Lab (mivielab), ViCubeLab, Universiti Teknologi Malaysia, 81310 Johor, Malaysia
[2] School of Computing, Faculty of Engineering, Universiti Teknologi Malaysia, 81310 Johor, Malaysia

* fazliaty.edora@gmail.com

Abstract. Extended Reality (xR) encapsulates various computer-altered reality technologies that cover virtual reality (VR), augmented reality (AR), and mixed reality (MR). xR is a technology that merges a virtual element into the real-world with the aims to enhance reality on the virtual world immerse onto real-world space. xR has been improved from time to time as the advanced immersive technologies to extend the reality we experience by either combining the virtual and real worlds or by generating a fully immersive experience. Remote collaboration in xR is a challenge since both parties need to have the same system and to set to parallel to xR environments. In a collaborative interface context, the user can be in a remote collaboration or face-to-face to sense the immersive environment. Human teleportation is transferring a human from a local location to a remote location, where the reconstruction of a human appears in a realistic visual representation. However, creating a fully realistic representation of the human figure need a complex 3D reconstruction method. Therefore, the paper describes the human teleportation in the xR environment using the advanced RGB-D sensor devices. It explains the phases to develop the real human teleportation in xR. The paper also discusses the proposed collaborative xR system that has successfully actualized the user teleportation.

1. Introduction

As technology becomes advance, there is a lot of device far more significant that can replicate and imitate our real-world environment. These technologies called Extended Reality (xR). xR encapsulates various computer-altered reality technologies that cover virtual reality (VR), augmented reality (AR), and mixed reality (MR). MR can bring us into space with computer visual graphics. MR is a technology that merging of real and virtual worlds somewhere along the "Reality-Virtuality continuum," which connects authentic environments to completely virtual ones [1]. MR aims to enhance reality in the virtual world immerse in real-world space. There is much research on Extended Reality (xR), which in every way, makes the human lifestyle more present in the virtual world. In order to create an xR application, accurate positioning and tracking require aimed at aligning the virtual and real world. xR is different from VR in the way of user able to manipulate the object such as in the MR environment while able to compare with real-world space and more similar to AR. MR also make the user aware of the surrounding while not ignoring the surround. Therefore, this paper describes how the MR provides user best realism in virtual and real-world space better than AR and more presence in VR.

The human teleportation in the xR environment using the advanced RGB-D sensor devices [2], often referred to telepresence [3]. This paper explains the phases to develop the real human teleportation in xR. Collaborative xR interface allows users to perform cooperative works seamlessly where the content is a mixture of real and virtual worlds [4]. The paper also discusses the proposed collaborative xR system that has successfully actualized the 3D telepresence in an urban planning prototype.

2. Extended Reality (xR)

What sets MR and VR apart is the ability to be seen and present in a real-world environment. The user would feel separated from the real-world and traditional tool in an immersive VR [5]. MR environments are overlaying of virtual objects in the real world. It can be concluded that MR is allowing the interaction between users on the real-world while maintaining the virtual images through high bandwidth of communication between users and impulsive manipulation on the virtual world [6].

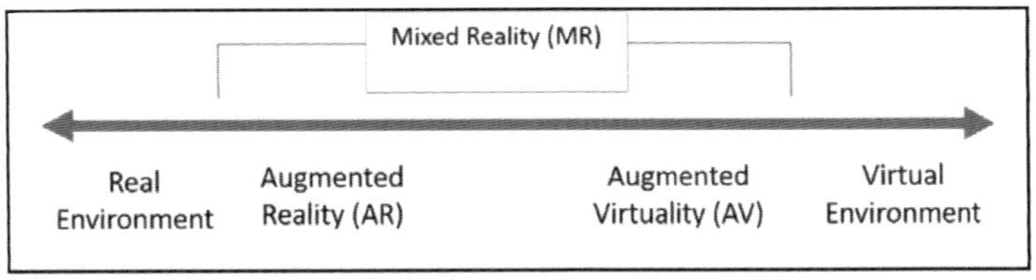

Figure 1. Reality-Virtuality Continuum [1].

From Figure 1, a classification from [1] on Reality-Virtuality Continuum shows that MR consists of AR, AV, and VR. AR is a real-time technology which registered in 3D with display technique. AR is 3D virtual objects which are integrated into a 3D real environment in real-time [7]. AR is referred to as an interface in which 3D computer graphics are overlaid over real objects through head-mounted or handheld displays [8]. The goal of AR is to create the insight that the virtual elements are a part of the real environment. AR is used to project the virtual elements layered on the real object by superimposing a correct view and position of the 3D model onto users' point of view of the real environment [8].

VR is an immersive computer-generated environment that locked the user inside the fabricated environment and cuts the user from the real-world environment. VR is a computer-simulated environment that can convince a human's brain about the existence of the computer-generated environment [9]. VR acts as a medium where it allows users to communicate and exchange with other users while the users immersed and present in the virtual world. The experience in VR is not limited to the fabricated environment only but can also include novels, movies, or any other environment. VR is the combination of display technology with a 3D model in real-time, providing an interactive environment allowing the user to manipulate and, at the same time, feel the immersion of the digital world. In contrast, VR changes the user's perspective of being in a new environment rather than being in a real environment. VR is an immersive and interactive technology using a joystick or mid-air interaction that replaces reality by replacing the real world with the simulated world.

3. Literature Review

3.1. Remote Collaboration Interactions

User collaboration in MR is a challenge since both parties need to have the same system and to set to parallel to MR environments. If the MR collaboration needs a specific space, both users need to replicate both places to have precise and accurate positioning in both users. Lack of co-presence feeling makes collaborative system absence of user experience [4]. A good user collaboration needs a good network architecture to achieve smooth data interchange [5].

A collaborative interface or remote collaboration system is to enable people who are far apart can sense in the same space [10]. The use of a collaborative interface has been utilized in the application to perform an assessment that requires multi-user. The collaborative interface allows its users to work cooperatively towards the shared task [8]. Communication is the most crucial aspect of any cooperative task. In a collaborative interface, visual aid is vital to navigating users around the system. The previous research in collaborative interface produces an issue of lack in social presence. Hence, the adoption of non-verbal communication cues can enhance a collaborative system relating to the issue of social presence [4]. In order to counter this issue, the proposed application include visual aid and avatar to enhance the collaborative interface.

The collaborative interface requires multi-user and network connection to perform the given task. In this research, a remote collaboration takes place between MR and VR users, in which each user use Head Mounted Displays (HMD) device. The use of HMD in MR environments creates the possibility of having a more immersive collaboration experience [9]. The issue may arise through the multi-user interaction as each user have a different platform and input to interact and manipulate in the shared network space [10]. Another remaining issue in the integration is the selection method; users need to select the shared objects at the same time. In order to overcome this issue, the authority method is used to determine which user has given a chance to manipulate the shared object. Using the authority method keeps minimum networking and ensuring a smooth collaborative interface.

3.2. 3D Telepresence

User teleportation can be represented as 3D telepresence [11]. 3D telepresence requires many processes. One of the most complex processes is a 3D reconstruction from input stereo or RGB-Depth camera [12]. Developing a real-world environment with high-fidelity in VR through 3D reconstruction is difficult. The teleportation of the user into the virtual environment requires many algorithms and equipment to compress and extract 3D data. The 3D telepresence gives a critical to the user experience in real-time performance [13]. A slight mistake in placing the camera and sensor would cause a flaw in data and can lead to inaccurate positioning in MR environments. User teleportation in xR environment is a challenge in this research since the process needs hardware that can enable and capture an object in 3D data into the application [12].

The standard camera can produce 2D RGB data, such as image and video. The output of a standard camera is insufficient in enabling and capturing an object in 3D data where the standard camera cannot capture depth data into the application. Therefore, a depth camera is a requirement in this research for stream and capture 3D object into the application [14]. There has been an issue of reconstructing 3D data into the application, in which the previous research leads to creating a low resolution of the 3D model [15]. The user teleportation process requires a loop procedure of 3D reconstruction by post-processing the captured object data and display it into the application. Since the output of user teleportation data is in 3D, the transferred data contains RGB and depth input of the captured object, which leads to a massive amount of data transferred. The remaining issue in user teleportation process is how to stream the post-processing of RGB and depth data over a network because of the vast data transferred required a high bandwidth of internet speed to overcome the issue [16].

Therefore, the proposed solution for the user teleportation issue is by the stream the RGB and depth data from the remote user by extending the connecting wire of the depth camera through the local user. Teleportation is also a common concept involving a distinct movement to a destination [17]. User teleportation is a method to capture 3D objects and teleport into virtual space. Early telepresence researches were focused on capturing dynamics scene using an array of cameras [3]. User teleportation gives the advantage of cutting the cost of travel and time spent to meet each other. In the early systems were developed, hardware and computational were the limitations at that time as the systems were able to capture low-resolution of 3D models and transmitted to the remote viewer. There were numerous researches focusing on 3D telepresence systems, and there were encountered difficulties for real-time capture, transmission, and displays [11]. From the past research, mostly sacrifice realism for faster performance in remote collaboration in telepresence.

3.3. 3D Reconstruction Method

Depth processing enables objects taken from different viewpoints to be backed up into 3D. Then, the registration and alignment of all points could completely restructure the environment. There are two ways to determine depth, first is through the stereo cameras, and the second is through the structured light sensors. User teleportation needs a stereo camera to provide depth data to capture users in the workspace. A stereo camera is a camera with two or more lenses together with an individual image sensor for each lens. The camera allows us to replicate human binocular vision and therefore delivers the ability to capture 3D images. Nowadays, the stereo camera is equipped with an Infrared (IR) sensor to improve the ability to capture the depth data. Recent research on user teleportation using multi-viewpoint to capture the user in the system [18]. The multi-viewpoint provides the presence of the user in the scene. Structured light is the most precise and fast in-depth estimation [19]. The most stereo camera provided structured light generate from the IR sensor. The stereo camera provides RGB-Depth (RGB-D) data from the lens sensor. RGB-D data is another advanced solution real-time depth estimation. Each sensor in the stereo camera produces an image with estimated depth by match and patch one image to another.

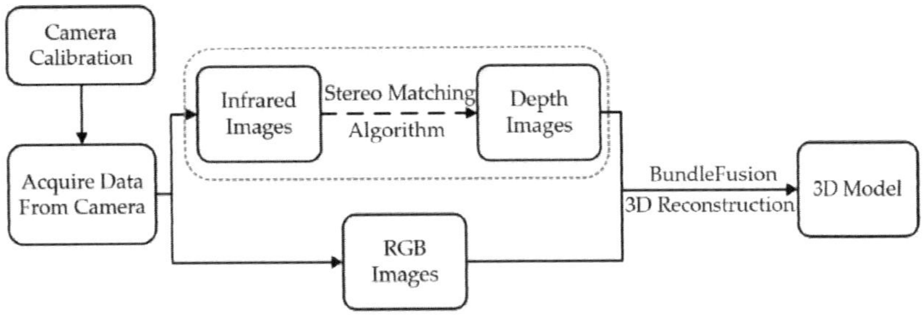

Figure 2. 3D Reconstruction Process using RGB-D camera [20].

The foreground segmentation stage follows the depth estimation. Foreground segmentation extracts the foreground object in an image by grading each image pixel as a foreground or background [3]. The output from depth estimation has created an image with various depths. Foreground segmentation eliminating the background and focuses on the data. This process as shown in Figure 2 plays an important stage in the later process of user teleportation, which is 3D reconstruction.

A modern stereo camera can make exhaustive scans of objects by generating point cloud data containing millions of points [21]. A point cloud is a set coordinate point which represents the surface of the data. One of the methods of 3D reconstruction is by visualizing the 3D model in the form of the point cloud. Point cloud has a problem in which is it lacks the connectivity structure of the underlying mesh and hence requires additional post-processing steps to extract 3D geometry from the model [22]. The output of point cloud may cause an error if not covers the most surface of the captured data as the stereo camera cannot capture the occluded from the camera viewpoint. Point cloud stitching is implemented to reconstruct the captured data by filling the hole

Point cloud data only consists of depth data and no colour data. In order to process a coloured 3D data, colour texturing is applied. Colour texturing approach is by computing the colour of each pixel by blending the RGB images according to its surface normal and the camera viewpoint direction. In colour texturing, depth estimation plays an important role in differentiating the colour between two surfaces. The missing sections of geometry might affect inaccurately projected colour to the surfaces behind it. Past research has implemented stitching and combining the nearest colour texture to overcome the missing colour texture in surface reconstruction [23].

IOP Conf. Series: Materials Science and Engineering **979** (2020) 012005 doi:10.1088/1757-899X/979/1/012005

4. Proposed xR Application

4.1. 3D Telepresence Input

The 3D telepresence process and method has been developed to enable user teleportation. This setup for the user teleportation required depth camera, an Intel RealSense D435, it was used to capture the RGB and depth data. The depth camera data undergo point cloud rendering process to display the data in 3D. This process is to map the input data from the depth camera into the output display. From this point cloud rendering, the proposed system captured the user's human body and process it into the application to achieve the user teleportation method.

4.2. xR Apparatuses Setting

In the xR setup, the display technology was required as the user needs to wear HMD to display output in it. The Project NorthStar HMD (PN-HMD) was used for the MR user and Oculus Rift HMD (OR-HMD) for the VR user. For PN-HMD, the gesture input was defined by Leap Motion Sensor for MR user interaction, and for OR-HMD, the Oculus Touch device has been used for VR user interaction. MR user has experienced the application using PN-HMD. The PN-HMD needs to attach along with the Leap Motion device to obtain the gesture input and Intel RealSense T265 on a dedicated mount to acquire the positional tracking. Lastly, the PN-HMD needs a positional tracking input to move around in the scene. Intel RealSense T265 be able to provide the Six Degrees of Freedom (6DOF) positional input.

Figure 3 shows the collaborative interface components overview, which includes all the hardware required to complete this application. Networking was crucial in this research. To integrate into the collaborative interface, interaction for spawning, delete, and manipulation of 3D objects was synchronized in a network. Performing integration and collaboration in the Unity3D game engine, take a plugin to establish a connection between the user. Unity3D engine process, the renderer, and the input handler of the game. In this phase, the Photon Unity Networking (PUN) plugin was used as a networking connection between users as referred to [8]. PUN server act as a host computer that handles all client connection requests maintains and manages client-shared virtual content, and processes data storage and retrieval.

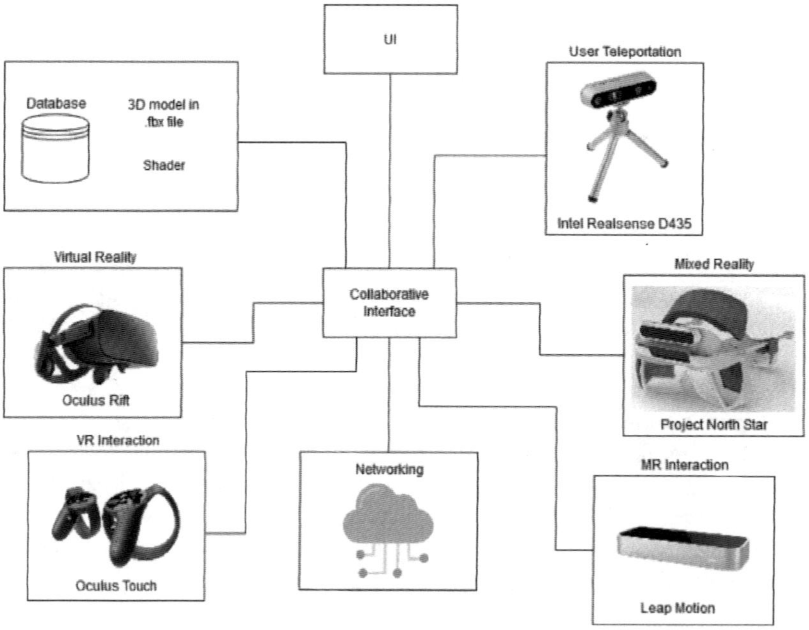

Figure 3. xR apparatuses and interaction in the collaborative interface.

4.3. Remote Collaboration Interactions

The prototype has been set up a shared environment for the user to experience the same cooperative task in a shared space. The urban planning task in this collaborative interface has several interactions in the shared space. The shared environment consists of a 3D object for urban planning tasks. The application consists of two clients assign as local users as Client 1 and remote users as Client 2. Local user must create room to start the application, and the remote user must enter the room. Both users experienced in a shared space to conduct urban planning activity. The shared room connected to Ethernet for synchronizing action. The data from the user input device when manipulating the object was sent to the server. The server sends the processed data back to the user to update the output in the user view. Figure 4 illustrates the process of collaborative interface application in the MR environment.

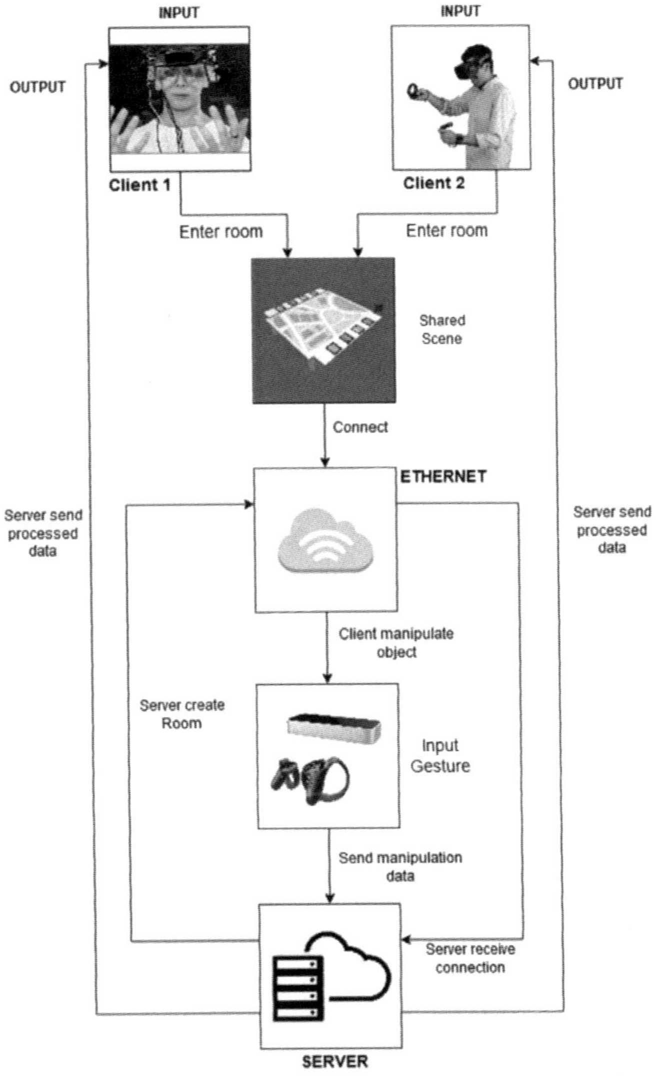

Figure 4. Remote Collaboration Interactions framework.

Both HMD using gesture input for the interaction. There are joystick and Leap Motion devices used in this research. The Leap Motion gesture interaction, gesture consists of drag, tap, pinch and release. Tap gestures are discrete to detect one or more fingers touching the 3D button. The fingers involved in these gestures must not move ominously from the initial touchpoints.

Tap gesture involves in this research is for tapping the 3D button for spawn interaction. A pinch gesture is a continuous gesture that tracks the distance between the first two fingers that touch the 3D object. Pinch gesture is used for grabbing the virtual object in the MR environment. Drag gesture is a motion that brings an object from one location to another. In this research, drag gesture is enabled along with pinch gesture for dragging a virtual object in a collaborative interface. Release gesture is a movement where all fingers are extended. Release gesture is subsequently of pinch gesture. This research uses a release gesture for release user virtual object from hand. Figure 5 shows the hand gesture interaction technique to perform drag, tap, pinch and release. The real hand gesture using Leap Motion has been learnt in [24], their work has been explored in this research.

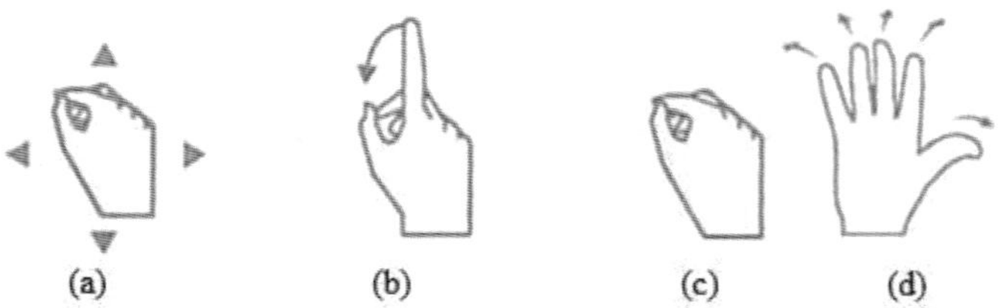

Figure 5. (a) drag (b) tap (c) pinch (d) release.

As both users have different input, the user performs different method to interact with the application, but both shared the same task, which spawns, delete and manipulate the 3D object. Therefore, the collaborative interface has a different executable program to respond to each user. Interaction for object manipulation shares the same process in MR and VR side.

The interaction involves consist of tap the 3D button and grab the virtual 3D object for translate and rotate manipulation. The rigid finger body from the input device is used to detect the collision between the 3D button and finger. If the rigid body is colliding, the object is spawned in shared space. Grabbing the 3D virtual object is performed by getting coordinate of index and thumb finger from input device for pinch gesture. Manipulation of a 3D virtual object can be performed if the object is pinch by the user. The manipulation involves in this research are translate and rotate the object. After performing the manipulation, the user either can repeat the interaction or end the process.

5. Implementation and Results

The user teleportation illustrates that the user teleportation process was achieved by executing the point cloud processing. Then, followed by the experimental setup of the remote user, the RGB-D camera was connected to the local PC. The results of the real-time user teleportation shown in Figure 6. Urban planning has been explored to implement the framework. Urban planning features and requirements have been referred to [25].

Figure 6. Real-time 3D Telepresence in Urban Planning prototype.

The collaborative interface has been developed simultaneously with the integration process. In general, both users experienced a synchronous update from each other. All the spawn object spawn at the network space; hence the manipulation also updates on the server, not on individually. The shared space of the collaborative interface is shown in figure 6. In the shared space, the UI only active for each side of the user.

Since the research is working with a remote collaborative interface, there were two setups, which are for the local and remote sites. The experimental setup for this research is shown in Figure 7 (a) and Figure 7 (b).

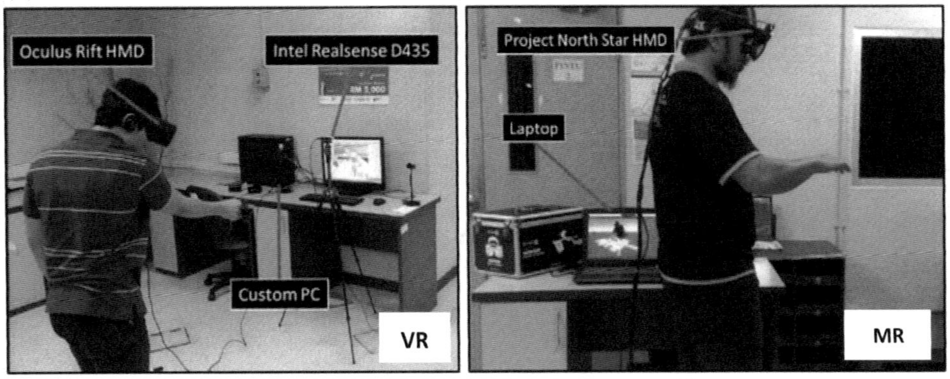

(a) (b)

Figure 7. Two users in the two workspace setups in real-spaces, remote site and local site.

The activity involves in the application are spawn, delete, and 3D manipulation of a 3D object. All interaction with the 3D button was user tap the button. Figure 8 shows the spawn object interaction, while Figure 9 shows the delete object interaction. Figure 10 demonstrates both users grabbing a virtual object. Figure 11 shows the collaborative interface output for VR and MR.

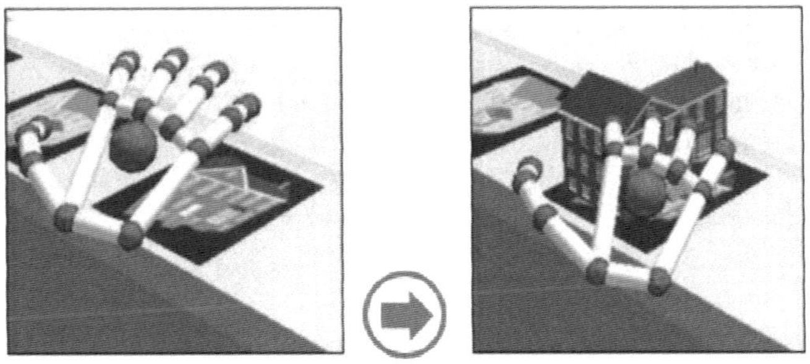

Before After

Figure 8. Spawning object.

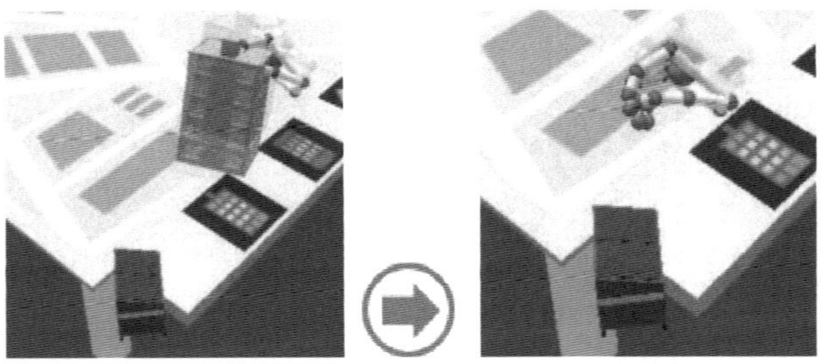

Before After

Figure 9. Deletion object.

IOP Conf. Series: Materials Science and Engineering **979** (2020) 012005 doi:10.1088/1757-899X/979/1/012005

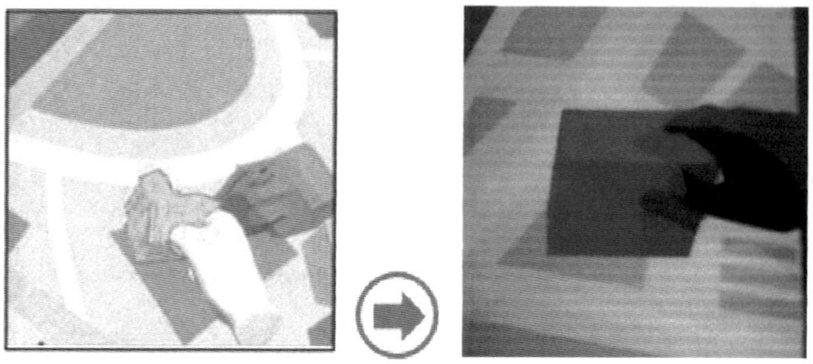

Remote site Local site

Figure 10. Translation and Rotation

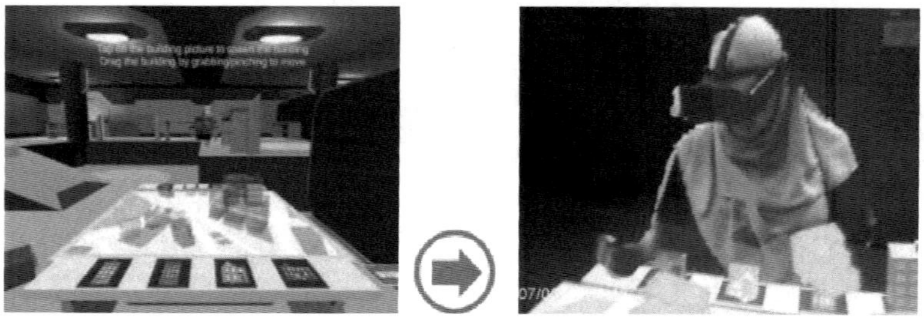

VR MR

Figure 11. Remote Collaboration Interactions

The xR application creates the synchronous update on 3D manipulation method in the collaborative interface; the 3D object is executed with the function that updates and retrieves the data from the server. All 3D object is involved with Photon Transform View and Photon View script to call the updated positional and rotational data in the server. The PUN component that links to the 3D object to update the manipulation data such as position and orientation coordinates. This proposed application had avoided the network congestion when an authority on the 3D object was given to the current user who has grabbed their desired object. If the 3D object was not assigned to the particular user ownership, the application requested from the other user for the ownership status. From this behaviour and condition, the manipulation of 3D object run in asynchronously. The process of requesting ownership between MR and VR user has started when the MR user is currently having the ownership of the 3D object. As soon as VR user grab the 3D, the ownership of the 3D object is transferred to VR user.

Table 1 shows the list of interaction command and each of the description. The interaction command exist in this research are select, spawn, translate, rotate, release and delete.

Table 1. Interactions and its descriptions.

Interaction	Description
Select	Using tap gesture for tapping the 3D button for activating the button.
Spawn	Instantiate 3D object above the 3D button position.
Translate and Rotate	Using pinch and drag gesture to manipulate the 3D objects by user hand movement.
Release	Using release gesture to discharge current holding 3D object in hand.
Delete	Destroy 3D object by using drag gesture on 3D object into delete 3D button.

6. Implementation and Results

This paper presents the phases to develop the real human teleportation in the xR environment using the advanced RGB-D sensor devices. The paper also discusses the literature reviews of user teleportation, which includes user teleportation method, process, and pre-defining setup. This paper also demonstrates the xR application and explains the development of collaborative interfaces and the experimental setup for the xR. The research acquired a user to wear HMDs for the VR and MR displays. For the collaborative interfaces, the proposed xR system runs the urban planning prototype to execute the cooperative task between two users with the intention of implementing 3D telepresence for remote collaboration. The contents are for urban planning; two users work in a shared environment. Issue in bandwidth network connection that used to connect users from different worlds into a single application has been highlighted. This study has study that distance measurement is one of the factors to affect the shared system. MR user is set as a local user, while the VR user becomes a remote user. The inputs have been captured for both users to interact with the MR environment. Finally, this paper ends with the user interactions and to determine the user interaction for both users, such as deletion, spawning the objects, translation and rotation. The integration of a 3D telepresence to transport the remote user into the shared environment has been successfully demonstrated in this study. The integration is using a network plugin to establish the connection between two users, and the stable internet connection is required to avoid transmission delay. The manipulation in the remote collaboration interface is synchronizing in the network. The proposed xR system has captured the user's human gesture and figure, and process it into the shared environment to complete the user teleportation or being known as 3D telepresence.

Acknowledgement
We would like to express our appreciation to Mixed and Virtual Reality Laboratory (mivielab), ViCubeLab at Universiti Teknologi Malaysia (UTM) for the facilities and technical supports. We also would like to thank Mr Noah Zerkin, who allowed me to have Project North Star HMD and lots of support from the Project North Star community.

References

[1] Milgram P and F Kishino 1994 A taxonomy of mixed reality visual displays. *IEICE TRANSACTIONS on Information and Systems* 77(12) p 1321-1329.
[2] Orts-Escolano S, Rhemann C, Fanello S, Chang W, Kowdle A, Degtyarev Y, Kim D, Davidson

PL, Khamis S, Dou M and Tankovich V 2016 Holoportation: Virtual 3d teleportation in real-time *In Proc. of the 29th Annual Symp. on User Interface Software and Technology 2016* pp 741-754

[3] Fadzli F E, Ismail A W, Aladin M Y F and Othman N Z S 2020 A Review of Mixed Reality Telepresence. *In IOP Conf. Series: Materials Science and Engineering* Vol 864, No 1, p 012081

[4] Irlitti A, Piumsomboon T, Jackson D, Thomas BH 2019 Conveying spatial awareness cues in xR collaborations *IEEE transactions on visualization and computer graphics* Vol 25(11) pp 3178-89

[5] Billinghurst M and H Kato 1999 Collaborative mixed reality *Proc. of the 1st Int. Symp. on Mixed Reality* p 261-284.

[6] Ismail AW and Sunar M S 2009 Survey on Collaborative AR for Multi-user in Urban Studies and Planning *Springer Berlin Heidelberg*

[7] Aladin M Y F and Ismail A W 2019 A review on multimodal interaction in Mixed Reality Environment *In IOP Conf. Series: Materials Science and Engineering* Vol 551, No 1, p 012049

[8] Nor' a M N A and Ismail AW 2019 Integrating Virtual Reality and Augmented Reality in a Collaborative User Interface *International Journal of Innovative Computing* 9(2).

[9] Craig A, Sherman W and Will J 2009 Developing Virtual Reality Applications: Foundations of Effective Design

[10] Nor'a M N A and Ismail A W 2019 A review on collaborative learning environment across virtual and augmented reality technology *In IOP Conf. Series: Materials Science and Engineering* Vol 551 No 1 p 012050

[11] Zhao M 2018 Towards high-quality 3D telepresence with commodity RGBD camera (Doctoral dissertation).

[12] Fuchs H, State A and Bazin J C 2014 Immersive 3D Telepresence Vol 47 pp 46-52

[13] Córdova-Esparza D M, Terven J R, Jiménez-Hernández H, Herrera-Navarro A, Vázquez-Cervantes A and García-Huerta J M 2019 Low-bandwidth 3D visual telepresence system *Multimedia Tools and Applications* Vol 78(15) pp 21273-90

[14] Manolova A, Neshov N, Tonchev K, Koleva P and Poulkov V 2019 Challenges for real time long distance holoportation to enable human bond communication. *In 2019 42nd Int.Conf.on Telecommunications and Signal Processing (TSP)* pp 529-534

IOP Conf. Series: Materials Science and Engineering **979** (2020) 012006 doi:10.1088/1757-899X/979/1/012006

Needs analysis for the development of interactive virtual reality-based educational media on combustion engine mechanical technology

Sutiman[1*], A Yudantoko[1], A Yudianto[1], M Solikin[1], I W Adiyasa[1], N M Suaib[2,3], F Sholichin[2], D A Irawati[2]

[1] Automotive Engineering Education Department, Universitas Negeri Yogyakarta, Colombo Street 1, Yogyakarta, Indonesia.
[2] School of Computing, Universiti Teknologi Malaysia, 81310 UTM Skudai, Johor, Malaysia.
[3] UTM Big Data Center, Ibnu Sina Institute of Scientific and Industrial Research, Universiti Teknologi Malaysia, 81310 UTM Skudai, Johor, Malaysia.

*sutiman@uny.ac.id

Abstract. Needs analysis is the first and necessary step for the researchers to design a newly developed learning media to find out a set of specific requirements to conduct the research activities. The study aims (1) to investigate the user's learning preference, (2) to select the suitable learning materials and to choose the appropriate type of engines, and (3) to identify the needs for the virtual reality development. A focus group discussion (FGD) comprising some representative users and experts in learning media has been involved in the activities with the use of the FGD guidance sheet as the supplementary tools. Additionally, the literature study has also been conducted to collect the necessary additional data to perform a 3D design of the engine. The data were analyzed qualitatively by examining the comment from the users, the experts, and the outcome of the literature review study. It was then concluded that (1) the users require the learning process, which does not need to alternate the learning objects. (2) The two-stroke single-cylinder engine was decided to choose by considering the complexity of the engine mechanical construction. Finally, (3) the needs for engine assembly drawing and virtual reality development have also been described.

1. Introduction

Practical lessons are necessary for students in the engineering field of vocational education to master several skills related to the area of study. However, some parts of the equipment to conduct the lesson are costly and possibly risky for students to use. The emerging technology in virtual reality (VR) has great potential to be implemented in the learning environment to overcome the cost and the risk problem in the vocational engineering education field. The utilization of VR has widely implemented many others fields such as medical training [1], military training [2], automotive and aerospace design [3], product design [4], entertainment [5] and engineering education [6]. The already implemented VR for learning environment proofs that these applications can be means of enhancing motivation and stimulating student's understanding of certain events [7], especially those students who cannot follow the conventional way of using the instructional materials for learning.

The field of combustion engine mechanical technology is chosen in this study, considering that the skills in mastering the subject related to the combustion engine are one of the cores of the learning activities. Considering the fact that the different technologies can be fully utilized in a VR system, it is expected that learning via a VR system will enable students to interact with a virtual scenario in a multi-sensory manner [8], the difficulties of the students to assembly the mechanical components for the real activities could be possibly replaced and benefits in better understanding of the practical lesson. The fully digitalized system of the technology allows the students to eliminate the risky interaction with heavy mechanical parts of the combustion engine and removing the limitation of the used space. Moreover, the study for the affordance in utilizing VR for learning and teaching achieved positive effects in supporting pedagogical aspects [9]. The mentioned capability of the VR system has a promising probability of creating interactive learning experiences in the topic of combustion engine mechanical technology as compared to contemporary educational media settings.

The requirements identification to conduct research in question is an essential activity to perform at the beginning stage of the research. Several previous studies conducting similar activities demonstrate that the need analysis is an inseparable activity in conducting the research. In this phase, the review of the VR utilization for learning will be conducted to review the types of VR used for training, the theoretical framework for a VR environment, and together with the environmental design for this technology in the learning environment [10]. The need for a VR simulator has been reviewed before developing the real equipment in dental education. The overview of the simulators and the link between VR simulation and current pedagogical knowledge was clearly identified [11]. A previous related study in determining the need for a virtual laboratory, according to the chemistry teachers, was conducted before developing the real equipment [11]. The results indicated that the majority of the respondents require this kind of integrated virtual chemistry laboratory in hybrid learning. The identification of the appropriate topic for the implementation of this technology was also identified. However, this work mainly analyzes the need to develop an interactive VR-based educational media for the subject related to the assembly process of a combustion engine mechanical technology. The project will be implemented for students in the subject area of engineering for vocational education as one of the topics of the practical lesson.

The main research problem in this study is how to develop an educational media based on VR technology to be implemented on combustion engine mechanical technology. Therefore, the needs analysis to develop the VR-based educational media is necessary to perform. The aims of the study are to investigate the student's learning preference, to select the most appropriate learning materials, and to identify the needs of engine drawing to be implemented in the VR technology. This work will essentially contribute as the initial pace to develop the VR-based educational media and properly identify the requirements to conduct the subsequent research.

2. Research Method

The ADDIE model was employed, comprised of Analysis, Design. Development, Implementation, and Evaluation phases [12] to develop the entire system of VR-based educational media. This work was the first step of that model, which was the Analysis Phase. In this phase, there are three kinds of analysis that had been done which were student learning preference analysis, learning materials analysis, and needs of VR development analysis. This research adopted a descriptive study with a qualitative approach to analyze the results through the needs analysis.

The research respondents involved amounted to 56 (Table 1). The involved respondents were chosen to represent the students, lectures, and some experts in the required field. Questionnaire and Focus Group Discussion (FGD) were used in this work as a data collection technique, in which the questionnaire sheet and FGD guidance sheet were used as instruments in this work. The data from the questionnaire was analyzed quantitatively and the data from FGD was analyzed qualitatively.

IOP Conf. Series: Materials Science and Engineering **979** (2020) 012006 doi:10.1088/1757-899X/979/1/012006

Table 1. List of Respondents.

No	Respondents	Number
1	Automotive Vocational Students	40
2	Automotive Vocational Lecturers	4
3	Automotive Vocational Experts	2
4	Educational Media Experts	2
5	Instructional Design Experts	2
6	Automotive Industry Practitioners	6

3. Result and Discussion

Educational media could support students during their learning activities. Their achievements, motivation, and also collaboration skills could be developed by using educational media [13]–[17]. In addition, educational media also help teachers to deliver learning materials during their teaching and learning processes [13], [18], [19]. Developing educational media could have a positive effect on teaching and learning processes. However, the development of the educational media has to accommodate the needs of the users, so that the benefits of the educational media can reach the targeted outcome. The identification learning preferences perceived by the users, the choice of the most appropriate learning material, and the needs for VR-development need to be conducted before building the entire VR-educational media system.

3.1. Student Learning Preference

It is indispensable to determine whether a particular teaching methodology positively or negatively affects students' satisfaction with the learning process [20]. As a matter of fact, students need to be supported in their learning processes, both theoretical and practical lessons. The identification of the students' learning preferences can help students to maximize their true potential and to become lifelong self-directed learners [21]. It is because of the data from the questionnaire and also the FGD hint that they need support by developing educational media to make them easy to understand and master the required skills. The results of the data collection are depicted in the following figures. The first identification aims to analyze the perceived level of difficulties of the subject. The data was collected by means of a questionnaire having the total number of respondents depicted in Table 1. Figure 1 shows the level of difficulties of the automotive mechanical technology subject. The vast majority of the respondents involved in the survey said that the difficulties level of automotive mechanical technology is difficult. These results are described by 48% of the respondents said that the difficulties level of the subject is high and 20% of them said that the subject is very high. However, only a small percentage of students involved in the survey feel that the subject is easy by the results in 2% of them say that the subject is easy. The other respondents said that the subject is rather easy.

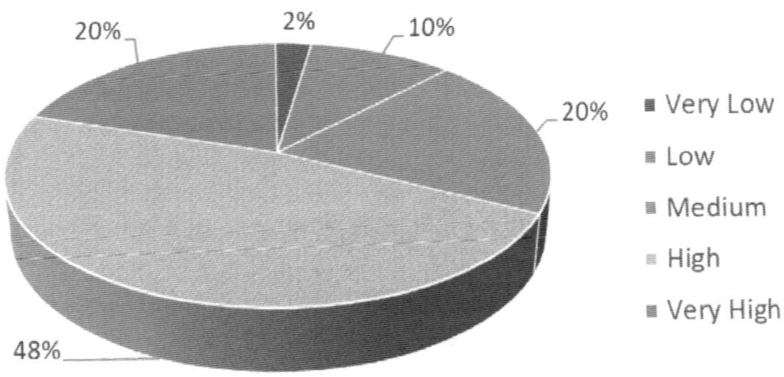

Figure 1. Level of Difficulties of Automotive Mechanical Technology Subject.

IOP Conf. Series: Materials Science and Engineering **979** (2020) 012006 doi:10.1088/1757-899X/979/1/012006

The interrelation between theoretical and practical lessons of a subject will altogether determine the quality of the learning and teaching process as well. Therefore, it is essential to identify the effectiveness of the educational media during the theoretical lesson of the related subject to the practical lesson under-identification. Figure 2 indicates the effectiveness of the educational media during the theoretical lesson of the subject related to combustion engine mechanical technology. 60% and 5% of the respondents indicate that the effectiveness of the educational media is respectively low and very low during the theoretical lesson. Approximately 22% and 13% of the students involved in the survey answer by medium and high. It can be derived that the effectiveness of the utilization of educational media needs to be improved.

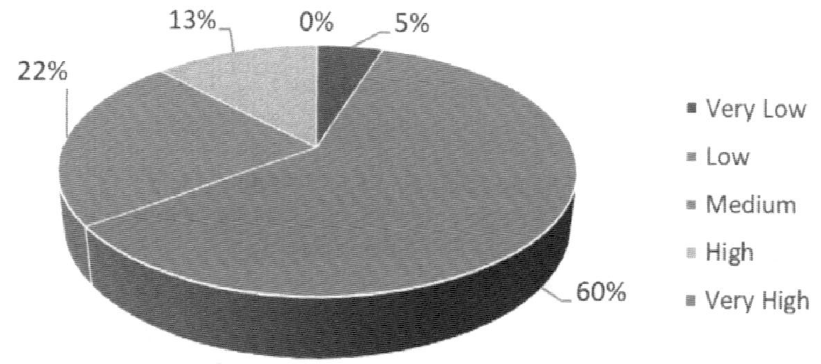

Figure 2. Level of Educational Media Effectiveness during Theoretical Lesson.

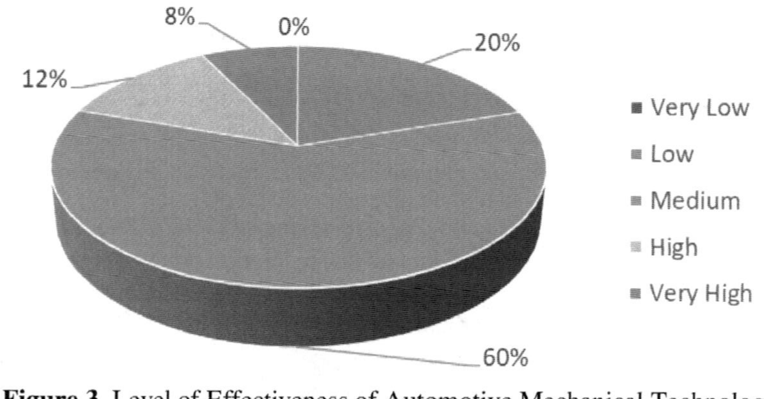

Figure 3. Level of Effectiveness of Automotive Mechanical Technology Practical Lesson.

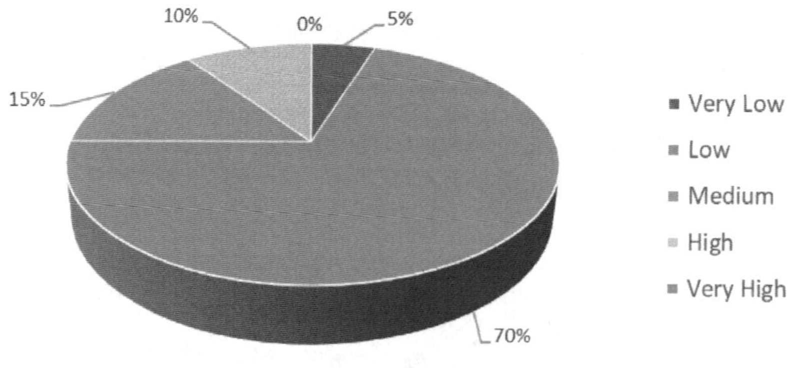

IOP Conf. Series: Materials Science and Engineering **979** (2020) 012006 doi:10.1088/1757-899X/979/1/012006

Figure 4. Level of Facility Availability during Automotive Mechanical Technology Practical Lesson.

Instead, Figure 3 depicts the effectiveness level of automotive mechanical technology practical lessons. More than half of the respondents answered with the medium level of effectiveness of the practical lesson. Only 12 % of them answered with high effectiveness, and the others are at a low level of effectiveness. Figure 4 describes the availability of the facility related to automotive mechanical technology practical lessons. The similar results indicate that almost three-quarters of the student involved in the survey said that the facility that is available to support the practical lesson of automotive mechanical technology is low.

Therefore, it can be derived from the results of the survey that there are many aspects that need to be improved in the teaching and learning process for automotive mechanical technology lessons. This subject is considered a difficult subject with the relatively low effectiveness of the theoretical lesson that is related to the practical lesson of the subject. Additionally, the practical lesson itself is perceived by the student to have a medium effectiveness level of the lesson activity. This problem might be related to the facility available to perform and support the practical lesson to be more accessible and understandable.

After answering the questionnaire, students were involved in FGD with lecturers that have three topics, which are a possible solution, educational media, and VR technology. The result says that VR technology could be as an educational media in order to be a possible solution for that problem. The students need an interactive educational media that use a state-of-the-art object and the one that is relatively new technology. Some already old educational media need to be replaced with the new one so that the effectiveness can be improved since the newly developed educational media with new technology is relatively more attractive for students. Modern educational media is also possible to improve the learning motivation for both theoretical and practical lessons in automotive mechanical technology. The students also require educational media that do not need to alternate the learning objects. The VR technology is finally preferable to be developed based on the mentioned considerations.

3.2. Suitable Learning Materials

The enormity of learning materials contributed to the struggle of choosing the most appropriate learning materials for a specific learning topics [22]. The FGD was used to find the most appropriate educational media materials to be chosen to develop the VR based educational media in this work. Vocational students, lecturers, automotive vocational experts, and automotive industry practitioners are involved in this FGD. The topic is to choose the kind of automotive mechanical component that needs to be developed. The answer is two-stroke engine mechanical components. There are some considerations in choosing this type of engine. Since this VR project is the initial project to be further developed for educational media in this work, the two-stroke engine is chosen by considering the level of complexity of the component. Therefore, two-stroke engine components are very considered appropriately to become a pilot project for VR technology as an educational media.

3.3. Needs for 3D Design in VR Development as an Educational Media

A VR system incorporates a set of technologies that allows users to experience and explore three-dimensional environments generated by computer [23]. It is necessary to identify the needs to build the virtual learning environment to be the educational media applied in the teaching and learning process. In this step, FGD was also used in this issue by involving educational media experts and instructional design experts. There are several results from this FGD. Firstly, in order to draw the engine components as materials to be developed in VR technology, the computer that has minimum RAM 8 GB and Graphic Card 4 GB is required. Secondly, Autodesk Inventor 2019 Student Version is chosen for this project. Thirdly, the reverse engineering method was chosen as the method to draw the engine mechanical component. In order to verify the drawing, the so-called motion and component interference conformity are used to ensure the drawing of the object. Other tools are also required to conduct the VR Development, such as vernier calipers, micrometers, cylinder bore gauge, and wrench set are required as measurement tools. After finishing the drawing of the object, The Unity software is the preferred platform to create the VR experiences for the educational media.

4. Conclusion

A conclusion can be derived that the needs analysis is an essential step for the development of VR educational media for automotive mechanical engine technology. The students require a learning and teaching process by employing state-of-the-art technology to facilitate students during theoretical and practical lessons. The learning object preferable by students is the one that does not need to alternate between the students and to maximize the utilization for students. The two-stroke single-cylinder engine is decided to be developed in the VR system. The need for VR development in terms of drawing and VR development system requires software and hardware related to component drawing and development of VR program.

Acknowledgments

Our sincere gratitude is expressed to Universiti Teknologi Malaysia as the international research partner, and to Universitas Negeri Yogyakarta for the funding of the research.

References

[1] I. J. A. Davidson *et al.*, "Virtual Reality: Emerging Role of Simulation Training in Vascular Access," *Semin. Nephrol.*, 2012, doi: 10.1016/j.semnephrol.2012.10.009.
[2] M. Zyda, "From visual simulation to virtual reality to games," *Computer (Long. Beach. Calif).*, 2005, doi: 10.1109/MC.2005.297.
[3] A. Gomes De Sá and G. Zachmann, "Virtual reality as a tool for verification of assembly and maintenance processes," *Comput. Graph.*, 1999, doi: 10.1016/S0097-8493(99)00047-3.
[4] P. Gaoliang, H. Yu, L. Xinhua, J. Yang, and X. He, "A desktop virtual reality-based integrated system for complex product maintainability design and verification," *Assem. Autom.*, 2010, doi: 10.1108/01445151011075799.
[5] C. Stapleton, C. Hughes, M. Moshell, P. Micikevicius, and M. Altman, "Applying mixed reality to entertainment," *Computer (Long. Beach. Calif).*, 2002, doi: 10.1109/MC.2002.1106186.
[6] A. G. Abulrub, A. Attridge, and M. A. Williams, "Virtual reality in engineering education: The future of creative learning," *Int. J. Emerg. Technol. Learn.*, 2011, doi: 10.3991/ijet.v6i4.1766.
[7] Z. Pan, A. D. Cheok, H. Yang, J. Zhu, and J. Shi, "Virtual reality and mixed reality for virtual learning environments," *Comput. Graph.*, 2006, doi: 10.1016/j.cag.2005.10.004.
[8] N. Gavish *et al.*, "Evaluating virtual reality and augmented reality training for industrial maintenance and assembly tasks," *Interact. Learn. Environ.*, 2015, doi: 10.1080/10494820.2013.815221.
[9] S. Minocha, A. D. Tudor, and S. Tilling, "Affordances of mobile virtual reality and their role in learning and teaching," in *HCI 2017: Digital Make Believe - Proceedings of the 31st International BCS Human Computer Interaction Conference, HCI 2017*, 2017, doi: 10.14236/ewic/HCI2017.44.
[10] E. A. L. Lee and K. W. Wong, "A review of using virtual reality for learning," in *Lecture Notes in Computer Science (including subseries Lecture Notes in Artificial Intelligence and Lecture Notes in Bioinformatics)*, 2008, doi: 10.1007/978-3-540-69744-2_18.
[11] E. Roy, M. M. Bakr, and R. George, "The need for virtual reality simulators in dental education: A review," *Saudi Dental Journal*. 2017, doi: 10.1016/j.sdentj.2017.02.001.
[12] N. Aldoobie, "ADDIE Model Analysis phase," *Am. Int. J. Contemp. Res.*, 2015.
[13] A. A. Sakat, M. Z. Mohd Zin, R. Muhamad, A. Ahmad, N. A. Ahmad, and M. A. Kasmo, "Educational technology media method in teaching and learning progress," *Am. J. Appl. Sci.*, 2012, doi: 10.3844/ajassp.2012.874.878.
[14] A. Bozanta and S. Mardikyan, "The effects of social media use on collaborative learning: A case of Turkey," *Turkish Online J. Distance Educ.*, 2017, doi: 10.17718/tojde.285719.
[15] R. Junco, G. Heiberger, and E. Loken, "The effect of Twitter on college student engagement

and grades," *J. Comput. Assist. Learn.*, 2011, doi: 10.1111/j.1365-2729.2010.00387.x.

[16] M. Liu, L. Horton, J. Olmanson, and P. Toprac, "A study of learning and motivation in a new media enriched environment for middle school science," *Educ. Technol. Res. Dev.*, 2011, doi: 10.1007/s11423-011-9192-7.

[17] Y. Cao, H. Ajjan, and P. Hong, "Using social media applications for educational outcomes in college teaching: A structural equation analysis," *Br. J. Educ. Technol.*, 2013, doi: 10.1111/bjet.12066.

[18] A. Delić-Zimić and N. Gadžo, "Implementation of ICT in Education," in *Lecture Notes in Networks and Systems*, 2018.

[19] F. J. Palacios, M. E. Gómez, and C. A. Huertas, "Digital and media competences: Key competences for EFL teachers," *Teach. English with Technol.*, 2020.

[20] R. J. Murphy, S. A. Gray, S. R. Straja, and M. C. Bogert, "Student Learning Preferences and Teaching Implications," *J. Dent. Educ.*, 2004, doi: 10.1002/j.0022-0337.2004.68.8.tb03835.x.

[21] P. Kharb, P. P. Samanta, M. Jindal, and V. Singh, "The learning styles and the preferred teaching-learning strategies of first year medical students," *J. Clin. Diagnostic Res.*, 2013, doi: 10.7860/JCDR/2013/5809.3090.

[22] K. I. Ghauth and N. A. Abdullah, "Learning materials recommendation using good learners' ratings and content-based filtering," *Educ. Technol. Res. Dev.*, 2010, doi: 10.1007/s11423-010-9155-4.

[23] S. V. G. Cobb, "Virtual environments supporting learning and communication in special needs education," *Top. Lang. Disord.*, 2007, doi: 10.1097/01.TLD.0000285356.95426.3b.

Educational mobile game for learning English words

B K Ng[1]*, N M Suaib[1,2], A J Sihes[3], A Ali[2] and Z A Shah[2]

[1]School of Computing, Faculty of Engineering, Universiti Teknologi Malaysia, 81310 UTM Skudai, Johor, Malaysia.
[2]UTM Big Data Center, Ibnu Sina Institute of Scientific and Industrial Research, Universiti Teknologi Malaysia, 81310 UTM Skudai, Johor, Malaysia.
[3]School of Education, Faculty of Social Sciences and Humanities, Universiti Teknologi Malaysia, 81310 UTM Skudai, Johor, Malaysia.

*nebk96@gmail.com

Abstract. English is a second language in Malaysia and English language skills are vital for Malaysians. Mobile technologies such as mobile phone can be a great tool for educational purposes since it can increase students' engagement and is a popular tool for interactive digital media. This paper discusses the development of an educational mobile game called EWORD for learning English words, for primary school kids. There were four phases involved; preliminary analysis, game design, game development and game evaluation. In the first phase, the fundamental knowledge of English words and vocabulary, English language learning methods in Malaysia, traditional learning methods, and educational mobile game were studied to acquire adequate information about the requirements for educational mobile game. User Interface (UI), gameplay, and game mechanics were designed in the second phase, while the mobile game prototype was developed in the third phase. In the fourth phase, the evaluation was conducted to evaluate the usability of the educational mobile game prototype. Based on the results, the educational mobile game helps in boosting the learnability of English words among students and is capable to increase student engagement.

1. Introduction

English is the most-spoken language in the world and it is important for younger generation in Malaysia to learn and have a good grasp of the English language. The limitations of traditional classroom learning methods such as limited time and place for learning English, as well as the passive learning environment lead to the lack of motivation and interest of primary school kids to acquire the English language. A mobile game is a digitally based game that plays on a mobile device such as smartphone, tablet, Personal Digital Assistant (PDA), and so on [1]. Mobile technology such as mobile phone can become a great tool for learning English words through the mobile game as it can reduce the problems of traditional classroom learning methods such as boring and passive learning environment. By combining mobile technology with the educational aspect, it can form an educational mobile game that can be introduced as a new learning method and make the learning process interactive and fun but also focus on the educational elements.

Game-based learning can provide motivation for learners through interactive content and competitive presented from game compare to traditional learning methods [2]. An educational mobile game can be a new learning method for students to learn independently and acquire knowledge by playing this mobile game. Students can focus more on playing a game rather than reading books. This is due to the fact that students are free to explore and figure out the game strategies to accomplish the game mission that enhances the students' engagement. Players can learn and gain knowledge from the subject through the educational mobile game. In an indirect way, players are trained to improve critical thinking and problem-solving skills as players need to accomplish the mission to get scores.

2. Literature Review

2.1 English Words

A word means a single distinct meaningful element of speech or writing that can be combined to form a sentence [3]. English word can be defined as a word that can be spoken and written in the English language. A list of words known of a language can be defined as vocabulary. Vocabulary learning is essential for learners to gain vocabulary knowledge by understanding the meaning of new English words. Knowledge of vocabulary plays important roles to master language skills such as listening, speaking, writing, and reading.

2.2 English Language Education in Primary Schools Malaysia

English is important as a second language in Malaysia. Malaysia provides an English education curriculum for Malaysians since at preschool. Based on the Malaysia Educational System, the English language is a compulsory subject for students that studies at primary (6-12 years old), secondary school (13-17 years old) and post-secondary and tertiary levels of education (above 17 years old) but not compulsory for students to pass the standard national examinations [4]. Malaysia implements new Malaysia Education Blueprint (2013-2025) launched in 2012 and had introduced Primary School Standards-Based Curriculum (KSSR) for primary education as a new initiative by the Ministry of Education of Malaysia start from 2013 to enhance the English language skills of students to minimum proficiency levels through educational contents compared to New Primary School Curriculum (KBSR).

Standard-Based English Language Curriculum (SBELC) is introduced as a new English curriculum that implements in primary schools along with KSSR [5]. The implementation of SBELC aims to improve the English language proficiency of students from urban and rural areas. The SBELC emphasizes student-centered teaching as an approach for formal English language teaching and learning in classrooms. Student-centered teaching and learning approach more concentrate on students to foster critical thinking skills and problem-solving skills through a variety of activities [4].

2.3 Traditional Classroom Learning Methods

Traditional classroom instruction is a learning method with space restrictions such as classrooms, schools, and other locations [6]. There are two approaches of traditional classroom learning methods that are teacher-centered and student-centered. The teacher-centered approach is one of the traditional learning methods that emphasizes the role of the teacher to deliver the educational content by using textbooks as resources for the curriculum to teach students in a classroom [6]. In the teaching process, the teacher as the center and teach the students based on the teaching content in particular English textbooks with a fixed syllabus. However, students are passive in the classroom as students become listeners in the teaching process [7]. Therefore, students may lose interest in learning if unable to follow up on the topics taught by teachers as each student has different learning progress.

Whereas, student-centered approach is another learning method that focuses on students and allows students to work in a small group to complete activities to enhance communication skills and provide thinking space for students [6]. For example, students communicate in English, discuss English exercises together, give suggestions, and learn from each other. This approach is more flexible as it encourages the students to become active participants in the learning process compared to the teacher-centered approach.

However, traditional classroom learning methods provide limited opportunities as students learn and speak English in the classroom only with a maximum of an hour [8]. Besides, students need to attend the class in school on time. In the passive environment, students have to pay attention in class to follow up and absorb the content delivered by the teacher in a short time. Moreover, the teacher has limitation to focus on every student in class, while each student has different learning progress and proficiency level in English. In view of that, students may lose confidence as unable to follow up on the topics and incapable perform well in class like other students. Furthermore, there is a lack of motivation in learning English from textbooks as students feel boring to memorize the textbook.

2.4 Educational Mobile Game

A mobile game is a game that can be played via mobile phones and developed in diverse mobile handsets such as the Android operating system and Microsoft Operating System [9]. Education is the process of receiving or giving systematic instruction, especially at a school or university [3]. Therefore, an educational mobile game can be defined as a mobile game used for educational purposes.

Mobile phones are suitable to be utilized for educational purposes such as learning tools due to its extreme portability and convenience. Mobile game-based learning can be offered as a new learning method with the use of educational mobile games or apps. This learning method takes advantage of the mobility of mobile phones and time flexibility as the learner can access educational resources anytime and anyplace [10]. An educational mobile game is also known as instructional games designed to meet educational objectives through rules and constraints contained in the mobile game to achieve the goal of the mission [11]. Additionally, educational mobile games can be employed as mobile learning tools that enable users to learn educational content interactively and engagingly. The mobility and information accessibility of mobile technology play significant roles to enhance English language teaching and learning [10]. Hence, an educational mobile game can be a great learning tool for children to learn a second language such as English.

Based on the research done by [12], the result stated that students able to learn English words indirectly through playing education mobile games. It creates the learning environment for students to learn English words without particular intention as students need to identify the meanings of the words. The result of the research also shows that mobile games able to enhance the English vocabulary acquisition of learners. Moreover, there are several requirements of educational mobile game are determined as shown in Table 1 based on papers and research reviewed to enhance user experience in learning English.

Table 1. Requirements of Educational Mobile Game.

Requirement	Description
Characters and Storyline	The storyline of a mobile game that represents the scenarios of the game that applied for players to understand the core of the game. The characters in the mobile game should be well-designed and attractive as it is one of the major elements to enhance the learning experience of learners [1].
Rule	A rule-based mobile game that utilizes for learning purposes able to boost the motivation of learners [2]. This is due to the players have been motivated in thinking and using an appropriate strategy in the game to get a higher score by reading the rules of the game.
Immediate feedback	Immediate feedback displays in the game to the players that reflect the performance of the players in the game mission whether players have done it correctly or incorrectly [13].
Goal	Each educational mobile game should set a clear goal for the player to achieve [14]. It can increase the motivation of players and being concentrate on accomplishing tasks at each game level to score a success in the mobile game.
Reward	A reward is also one of the requirements of educational mobile games that enable the player to earn things after accomplishing tasks. This enables players to gain a sense of accomplishment after obtaining the rewards from game missions such as points, money, and trophies [13].

2.4.1 Current Related Educational Mobile Games or Applications

There are three existing educational mobile applications or games used for learning English that are Super Word King, Duolingo, and Memrise are reviewed in this paper. Based on the Table 2, it shows that three educational mobile game applications use different methods to teach language and focus on vocabulary learning Besides, these mobile games have several similarities in integrating game elements such as goals, levels, and rank with educational content that offer an interactive and engaging learning method for language learners. It also provides motivation for learners to achieve goals such as lessons in each level. In addition, most of these educational mobile games provide several useful features such as words review that can help to reinforce the language skills of learners. In short, an educational mobile game can be a great tool for language learners to learn a language in a fun and interesting way.

However, there are some limitations of these educational mobile games and applications. For example, all educational mobile games and applications mentioned above provide language learning in a bilingual setting that using one language to teach another language. In view of that, the user should know another language if wants to learn English. For instance, Super Word King more benefits to Mandarin speakers as it teaches English words by providing the instruction and meaning of English words in Mandarin. But it is limited to Mandarin speakers only for learning English words. On the other hand, Duolingo also using the bilingual mode to teach users to acquire a language. However, Duolingo doesn't provide many options for Asian speakers such as Malay speakers to learn English. Whereas, users of Memrise need to pay an extra fee for upgrading the application to Memrise pro as it provides additional features such as unlimited access to all game modes. In view of that, the user may unable to enjoy all the features during learning languages. In view of that, these mobile applications and games may not suitable for all primary school kids in Malaysia as different races and languages are spoken in Malaysia.

Table 2. Comparison between three educational mobile games that are Super Word King, Duolingo, and Memrise.

	Super Word King	**Duolingo**	**Memrise**
Mobile Operating Systems	Android, iOS	Android, iOS	Android, iOS
Learning Contents	Vocabulary	Vocabulary, Grammar	Vocabulary, Grammar
Learning Modes	-Provide vocabulary learning through listening, spelling, and meaning matching methods -Learning through levels of accomplishment -Provide the pronunciation guide of a native speaker	-Learn to pronounce the words correctly -Word meaning matching -Translate the sentence provided - Learning through levels of accomplishment -Provide the pronunciation guide of a native speaker	-Provide the pronunciation guide of a native speaker - Provide vocabulary learning through listening and words meaning matching methods
Features	-Words Review -Quick Challenges with different categories of English words -Performance Ranking	-Words Review -Daily Goal setting -Daily Practice Reminders -Variety of Languages Courses provided	-Speed Review -Daily Goal setting -Performance Ranking -Variety of Languages Courses provided
Limitations	-Limited to Mandarin speakers only for learning English words	-Malay speaker is not included as the options for learning English.	-Extra fee for additional features

IOP Conf. Series: Materials Science and Engineering **979** (2020) 012007 doi:10.1088/1757-899X/979/1/012007

3. Usability Testing

Usability Testing is a useful testing methodology to test and evaluate the usability aspects of an application. It is vital in improving the quality of user experience as usability testing emphasizes aspects such as ease of use, efficiency, and satisfaction. Based on the research done by [15], it revealed that usability testing contained 5 quality components that are learnability, efficiency, memorability, errors, and satisfaction. Table 3 shows the five quality components with the corresponding description.

Table 3. Five Quality Components.

Quality Component	Description
Learnability	To determine the ease of use of the game for the user to accomplish the task or mission while the user first attempts to the game.
Efficiency	A factor that focuses on the time for the user to complete the task given after the user had the experience with the game
Memorability	Concerns on the skills of a user to play the game during reattempting the game after a long time not using it.
Errors	Criteria to measure the errors made by the user and the ease for the user to fix it
Satisfaction.	To measure the level of satisfaction by a user after using the game.

4. Methodology

This section explains the methodology used in this paper. Figure 1 shows the flowchart of methodology that represents the procedures required to guide the educational mobile game development.

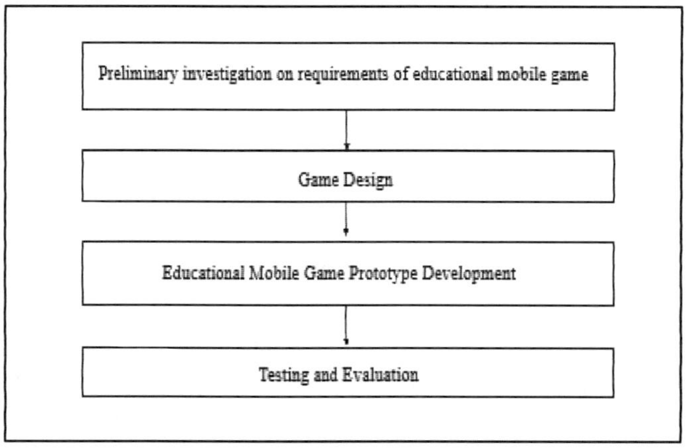

Figure 1. Flowchart of Methodology.

In Phase 1, the studies on English language education in primary schools of Malaysia and the traditional learning methods are conducted to collect the related information about the current learning method for the English language applied in primary schools. Besides, the research is conducted to investigate the existing educational mobile game and the technology that can be used to develop the game prototype to determine the requirements for educational mobile game development. Based on the research and interview, the requirements of educational mobile game such as goals, feedback, and reward can be applied in the educational mobile game to enhance the learning experience. After the

IOP Conf. Series: Materials Science and Engineering **979** (2020) 012007 doi:10.1088/1757-899X/979/1/012007

information collected in the previous phase, the game concept and gameplay of the educational mobile game are designed as well as game interfaces in Phase 2.

After the design phase is completed, the educational mobile game prototype is developed by using a suitable game engine and tools such as Unity and the Android Software Development Kit (SDK) to build the prototype in the Android platform in Phase 3. In Phase 4, the educational mobile game prototype developed is tested through user acceptance testing and evaluated by using usability testing to test the functionality and usability of an educational mobile game for primary school kids to learn English words.

5. Implementation

This section explains the implementation of this paper based on the methodology of this paper.

5.1 Educational Mobile Game Prototype Development

After the information collected in the previous phase, the game concept and gameplay of the educational mobile game are designed as well as game interfaces for game prototype development. For example, the number of levels and the total missions in each level. Each level is designed with three missions. Each English word is classified based on the categories and used in a suitable game level.

5.1.1 Curriculum Design

The educational content is a collection of English words according to the KSSR curriculum from Year 1 to Year 6 that utilized for the mission at each level. The mobile game is designed with three levels for the player to achieve. Each level has 3 missions. Each mission is packed with 10 English words under the same category. For example, level 1 is using English words that less than 5 letters, whereas level 2 and level 3 are using English words with 5 and above letters. Level 1 and level 2 provide each hint in the form of an image, while level 3 uses images and sentences as the hints for the player to complete the spelling of English words. The difficulty of the game is increasing in aspects of English words and hints provided if the player goes to a higher level. Players can learn and gain a list of English words with a different category in each mission by progressing through the levels.

5.1.2 Character and Rewards Design

The main character named Kiki in the mobile game is designed. The appearance of Kiki is designed as a beginner of magician and possessed magical objects such as a broom and magic wand. The rewards are the list of magical gems that give to a player after accomplished each mission. The player needs to collect all the magical gems to achieve the goal of the mobile game. Figure 2 shows the design of character and reward.

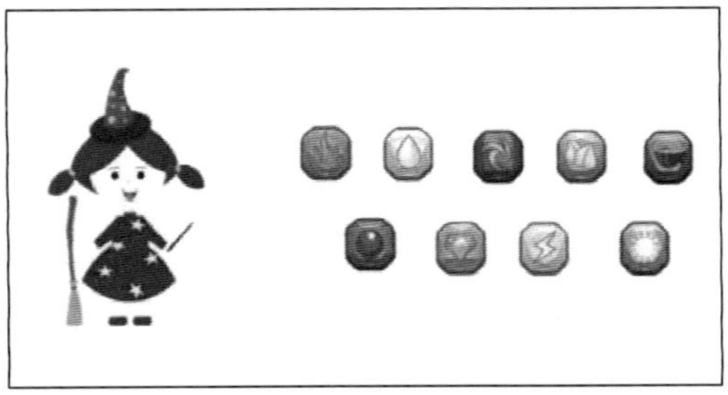

Figure 2. Character and Reward Design.

5.1.3 Interfaces Design

In this section, the interfaces of this paper are designed. A properly designed Graphical User Interface (GUI) of the game enable the player to start the game easily without knowing details before and provide the information. Figure 3 shows the interfaces of the game prototype.

(a)

(b)

(c)

(d)

Figure 3. Interfaces Design of Game Prototype that are (a) Start Page (b) Level Page (c) Mission Stage Page (d) Gameplay Page.

5.1.4Gameplay and Mechanics

EWORD starts with a story presented to the player with the theme of 'Magical' that describes the magical trial for the main character named Kiki to be a powerful magician by accomplishing the missions in each island to collect 9 magical gems. There are 3 levels in this game with the difficulty increasing while going to a higher level. Each level has 3 game missions and each game mission contains 10 questions for players. The mission starts with a question that providing an image or sentence as a reference for the player to complete the spelling of the English word. The question is picked randomly from the list of questions each time at all levels. The bubble objects are appeared and move randomly in the range of screen view. The player can tap on the desired position in the screen view to pop the bubble that is labeled with a letter. The player needs to find and tap on the targeted bubble to get the letter for completing the spelling of the English word. The target bubble pops and disappears after tapped by the player and the corresponding letter is filled in the English word. Moreover, the player requires to fill in the letters orderly according to the spelling of the English word. Players can get immediate feedback from the game such as pop the bubble that is labeled with the wrong letter through the role of sound effects and retry to answer the question in the given time.

There is a timer of 25 seconds shown in the game start to countdown for each question. The score in each mission is given based on the performance of the player in completing a list of English words correctly within the time given. For example, the player can earn 100 points if the player filled in all the letters of the English word within 5 seconds. If the timer runs out of time, the player is getting zero points and the game moves to the next question. The image of a new English word is given for the next question after the player completed the spelling of the previous English word. The player needs to answer 10 English words to complete the mission. The player needs to obtain a targeted score of 650 and above to get the magical gem as a reward. The game records and updates the highest score obtained by the player. Each game mission at each level are rated with stars based on the highest score obtained by the player. The list of English words used as 10 questions in the previous game mission is listed on the Review Page for the player to learn and practice the English words. The next mission is unlocked only if the player completed the previous mission. Whereas, a new level is unlocked after the player accomplished the previous level. The player can replay the mission to achieve the targeted goal if failed to obtain the targeted score in the mission. It provides the chance for the player to replay and review the

English words. The player can click the gem button on the level page to view the current gems collected. The goal of this game is to collect all nine magical gems for the game character to become a powerful magician by accomplishing all the levels.

5.2 Game Engines and Android SDK

Unity has provided a variety of important tools and functions to develop the educational mobile game prototype. First, unity allows users to build the project to the target platform such as Android. In order to develop a mobile game with the Android platform, it needs to add the Android SDK path to Unity that aims to allow Unity to create the Android Package Kit (APK) file and install it into a targeted mobile phone based on its Android version and its Application Programming Interface (API) level. Then, the unity project builds and runs as a mobile game on a targeted mobile phone. The educational mobile game prototype needs the function such as *Input.getMouseButtonDown* that can use to detect the position of the screen that tapped by the player to pop the target object on the tap position. The components such as collider 2D also need to add for the Two-Dimensional (2D) objects such as bubbles with the purpose of collisions detection. Due to the educational mobile game prototype is a 2D game, therefore, it can use the canvas that represents the screen view of the mobile game. All the User Interface (UI) elements are added as the child of the canvas such as text, images, button, and so on. It is useful for designing and building the interfaces of the mobile game.

6. Results

The result of this paper is an educational mobile game prototype, EWORD for primary school kids to learn English words. A user can interact with the content of the educational mobile game through the 2D touch screen input of the mobile device. The user can start the game with the first level. There are three levels in this game and each level contains 3 mission stages. The user needs to go through all the stages one by one and get target scores to unlock all the stages and levels. In each stage of a level, there are 10 English word questions prepared in the mission for the user to spell each English word in the correct order by popping the bubble labeled with the correct letter. The timer starts to countdown the time in 25 seconds for each English word question. After user completed all the questions, it shows the results with stars, scores obtained, and the highest score of the stage. The user can receive a magical gem as a reward if achieve the target score in the mission. Next, the review page is displayed with the list of English words used in the previous game mission for the user to learn the correct spelling of the English words and review it. If the user clicks on the Next button from the review page, it back to the current level page for the user to select the next stage to play. Once the user collects nine magical gems, it shows the final game win page to the user to indicate the user had finished and won the mission of all game stages of each level. The user can select the reset button to clear all the game data and record to replay all the game stages.

6.1 Evaluation

User acceptance testing and usability testing are conducted to test the functionality and user experience of the educational mobile game prototype developed in this paper. There are 3 target users aged between 10-12 years old and 10 non-target users aged between 23-30 years old are involved in user acceptance testing and usability testing.

6.1.1 User Acceptance Testing

Twelve actions that are the main tasks of the educational mobile game prototype are suggested for target users and non-target users to perform. All the results are recorded and tabulated as Table 4 shown below. Based on the results of user acceptance testing obtained, there are six out of twelve actions performed by all target users. Whereas, there are seven out of twelve actions that are performed by all non-target users. This is because these actions are the main tasks that needed to carry out by both groups of users to play and proceed with the game. Besides, the actions such as view the game tutorial, view the gem collection, and reset the game are not done by all target users and non-target users are due to the buttons are less noticeable and eye-catching for the users to select and able to continue the game although without doing these actions. Figure 4 shows the target user and non-target user tested the educational mobile game prototype.

Table 4. Results of User Acceptance Testing.

Action	Expected Result	Results of target users	Results of non-target users
Start Game	The game will start.	✓	✓
View story of game	User can understand the instruction given through story.	✗	✓
View game tutorial	User can understand how to play the game.	✗	✗
View gem collection	User can view the collection of gem collected.	✗	✗
Sound Setting	User can on or off the background music of game.	✗	✗
Reset Game	User can reset game data.	✗	✗
Select game level	User select the game level wanted.	✓	✓
Select game stage in game level	User can select the game stage in a level.	✓	✓
Tap on the letter bubble	User can tap the letter bubble to complete the spelling of English words.	✓	✓
Pause menu	User can select options such as back home, resume and restart the game.	✗	✗
Press OK to receive reward	User will collect a gem as reward	✓	✓
Review list of English words	User can view the list of English words in the game.	✓	✓

(a)

(b)

Figure 4. The respondents that tested the educational mobile game prototype are (a) target user (b) non-target user.

6.1.2 Usability Testing

The questionnaire is given for the respondent to fill after the respondent completed the time of 15 minutes in interacting with the educational mobile game prototype. The questionnaire contains two sections that are pre-test and post-test questionnaire. Based on data collected from the pre-test questionnaire, all the target users have no experience in using any educational mobile game to learn the English language. Whereas, only 3 out of 10 non-target users have the experience to use an educational mobile game to learn English.

Based on the results of target users as shown in Figure 5, it showed that educational mobile game is easy to use, memorable, efficient, and fulfilled with the other usability quality components. For example, the target users have been asked about the learnability of the educational mobile game prototype. There are 2 out of 3 target users choose "Agree" and 1 respondent chooses "Strongly Agree". Therefore, the result shows that most of the target users agreed that this game prototype was easy to use and learn English words. Moreover, most of the target users able to know the method to play with the gameplay of the game prototype and use the function of interfaces. The target users also agreed that the game prototype able to proceed smoothly on the mobile device and fewer errors occur during interacting with the game prototype.

From the results of non-target users as shown in Figure 6, most of the non-target strongly agreed that this game prototype was satisfying to use as 60% of non-target users choose "Strongly Agree" and 40% of non-target users choose "Agree" for the question related to the satisfaction. The non-target users felt satisfied to use the game prototype to learn English words. For the memorability of the game prototype, most of the non-target users able to understand the function of interface elements such as buttons and input controls as well as gameplay of mission at each level of the educational mobile game prototype.

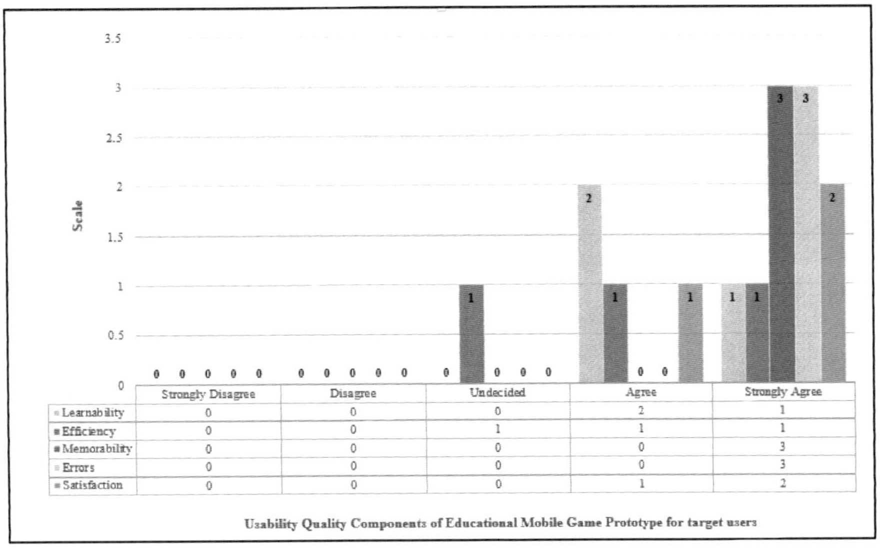

Figure 5. Feedbacks of target users to the questions related to usability quality components.

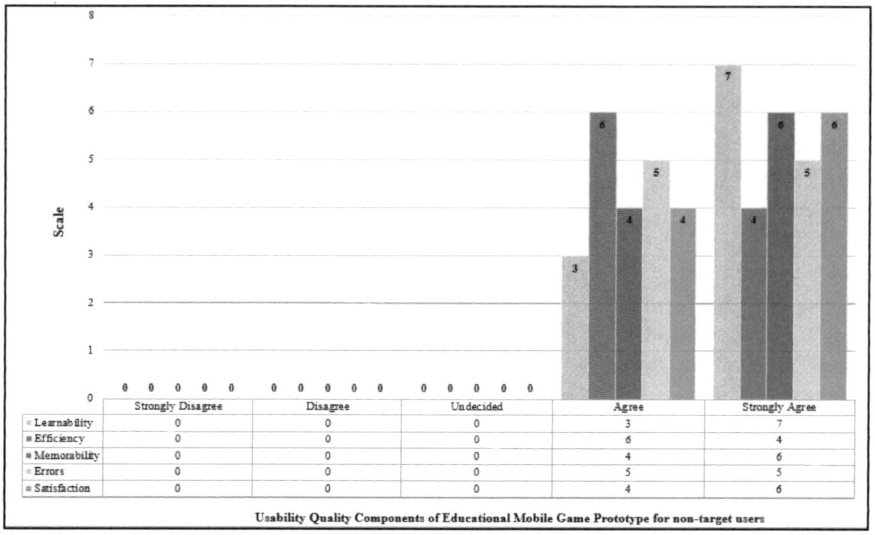

Figure 6. Feedbacks of non-target users to the questions related to the requirements of educational mobile game.

6.2 Evaluation Results

The results of usability testing were gathered and analyzed for the evaluation phase. The results obtained from the testing phase are important to further improve the functionality of the educational mobile game prototype for users to use and enhance user satisfaction. Both target users and non-target users agreed that this game prototype is targeted for primary school students to use as its learning content suitable and relevant to primary school standards of English subject. Besides, most of the respondents felt that the educational mobile game prototype is easy to use and can use it to learn English words. Most of the respondents also satisfied with the game prototype and found fewer errors during interacting with the interfaces of the game prototype. Moreover, the user experience of the educational mobile game prototype also is evaluated in the aspect of requirements of the educational mobile game applied such as rule, goal, immediate feedback, and rewards. Based on the results obtained of target users and non-target users, it show that the educational mobile game prototype is fulfilled with the requirements of educational mobile game.

7. Conclusion

EWORD is an educational mobile game prototype introduced as an option for a new learning method for primary school kids in learning English words. The requirements of educational mobile game were analyzed by collecting the knowledge and information from the research and studies that are important for the development of the educational mobile game prototype. The graphical user interface and game elements were designed based on the requirements as well as developing the educational mobile game prototype by using Unity3D. After the development phase, the testing and evaluation of this paper were conducted for improvements in the educational mobile game prototype. The usability of the educational mobile game prototype for primary school kids to learn English words was evaluated based on the results obtained. There are several limitations and improvements suggested for the future development of this paper. Due to the Movement Control Order (MCO) implemented in Malaysia from 18 March 2020, the situation prohibits any visit to schools or to invite primary school students as respondents for evaluation. Therefore, the testing process was carried out by involving two groups of respondents; 10 non-target users aged between 23-30 years old (for functionality test) and 3 target users aged between 10-12 years old (targeted users) as an alternative way to overcome current limitations. Therefore, we plan to do more testing such as involving more target users in testing, and to enhance this product by preparing more game levels for players to perform the more missions and learn a wider range of English vocabulary from the game, providing more game instructions, and adding more effects in aspects of audio and animation to provide feedback and to increase the enjoyment of the game.

Acknowledgments
The authors would like to express sincere appreciation to all people involved in this project and contributed to the success of this project.

References

[1] Shiratuddin N and Zaibon S B 2010 Mobile game-based learning with local content and appealing characters International Journal of Mobile Learning and Organisation 4 55-82
[2] Boeker M, Andel P, Vach W and Frankenschmidt A 2013 Game-based e-learning is more effective than a conventional instructional method: a randomized controlled trial with third-year medical students *PloS one* **8** e82328
[3] Stevenson A 2010 Oxford dictionary of English: Oxford University Press, USA)
[4] Azman H 2016 Implementation and challenges of English language education reform in Malaysian primary schools 3L: Language, Linguistics, Literature® 22
[5] Sulaiman S, Sulaiman T and Rahim S S A 2017 Teachers' Perceptions of the Standard-Based English Language Curriculum in Malaysian Primary Schools *International Journal of Instruction* **10** 195-208
[6] Relan A and Gillani B B 1997 Web-based instruction and the traditional classroom: Similarities and differences Web-based instruction 62 41-6
[7] Liu C and Long F 2014 The discussion of traditional teaching and multimedia teaching approach in college English teaching. In: *2014 International Conference on Management, Education and Social Science (ICMESS 2014)*: Atlantis Press)
[8] Ali M S 2003 English language teaching in primary schools: Policy and implementation concerns IPBA E-Journal 1-14
[9] Prasetya F 2015 Mobile Game Effectivenessfor Game Enthusiasts Who Have Little Spare Time to Play Games *SISFORMA* **2** 10-2
[10] Gangaiamaran R and Pasupathi M 2017 Review on use of mobile apps for language learning *International Journal of Applied Engineering Research* **12** 11242-51
[11] Gunter G A, Campbell L O, Braga J, Racilan M and Souza V V S 2016 Language learning apps or games: an investigation utilizing the RETAIN model *Revista Brasileira de Linguística Aplicada* **16** 209-35
[12] Ali Z and Ghazali M A I M 2016 Learning technical vocabulary through a mobile app: English language teachers' perspectives *International Journal of Language Education and Applied Linguistics*
[13] Aslan S 2016 Digital Educational Games: Methodologies for Development and Software Quality. Virginia Tech)
[14] Trisnadoli A, Muslim I and Novayani W 2016 Software Quality Requirement Analysis on Educational Mobile Game with Tourism Theme JSW 11 1250-7
[15] Nielsen J 2012 Usability 101: Introduction to usability (2012) URL: http://www. nngroup. com/articles/usability-101-introduction-to-usability/[Accessed November 2016] 9 35

IOP Conf. Series: Materials Science and Engineering **979** (2020) 012008 doi:10.1088/1757-899X/979/1/012008

Cultural heritage preservation efforts in Malaysia: A survey

N M Suaib[1,2*], N A F Ismail[2], S Sadimon[2] and Z Mohd Yunos[2]

[1]UTM Big Data Center, Ibnu Sina Institute of Scientific and Industrial Research, Universiti Teknologi Malaysia, 81310 UTM Skudai, Johor, Malaysia.
[2]School of Computing, Faculty of Engineering, Universiti Teknologi Malaysia, 81310 UTM Skudai, Johor, Malaysia.

*haida@utm.my

Abstract. Malaysia, just like its neighbouring countries in the region, has a rich and diverse culture and heritage treasures. What makes Malaysia more unique is its diversity as a multi-racial and multi-cultural country. These cultural heritages might become lost and extinct without any efforts in preserving and safeguarding due to modernization, assimilation, and globalization. We present an overview of different cultural heritage in Malaysia and available efforts to preserve these treasures found from literature. Digital preservation efforts that computer graphics, media scientists and practitioners could offer as alternatives in preservation of culture and heritage preservation will also be included in this paper.

1. Introduction

The need for cultural heritage preservation is certainly pressing, due to many factors such as development, modernization, climate change and assimilation. Due to its necessity, the United Nations Educational, Scientific and Cultural Organization (UNESCO) has established working committees and manuals to ensure that cultural heritage around the world receives proper attention and protection.

The context of cultural heritage relates to culture, heritage and cultural heritage. Culture has a broad meaning, but for clarification we refer to UNESCO's definition as the "complex whole that includes knowledge, beliefs, arts, morals, laws, customs and any other capabilities and habits" within a society. The World Heritage Convention in 1972 acknowledges that 'monuments, groups of buildings and sites' as heritage. However, as time progresses, interaction with humanity also has influence with the environment thus making the whole environment can be regarded as heritage.

There are different categories of heritage based on UNESCO's definition: cultural heritage, natural heritage and heritage in the event of armed conflict. We will outline different types of heritage in Malaysia in the next section, followed by efforts in cultural heritage preservation that were found in the literature. Highlight on the role of computer graphics technology in cultural heritage preservation will be presented before we conclude this paper.

2. Available Cultural Heritage in Malaysia

We will follow closely the categories outlined by UNESCO and the National Heritage Department, Ministry of Tourism and Culture Malaysia on various types of heritage available locally. Cultural heritage is diverse;

IOP Conf. Series: Materials Science and Engineering **979** (2020) 012008 doi:10.1088/1757-899X/979/1/012008

it is further divided into tangible and intangible cultural heritage. Tangible cultural heritage generally includes all things that can be observed such as things (movable – such as coins and paintings, and unmovable – such as sites and monuments) both on dry land and underwater. Intangible cultural heritage deals with a more subjective aspect such as tradition, dances and rituals. We listed some of the examples of cultural, natural and heritage in the event of armed conflict below – divided into those recognized in the UNESCO World Heritage list (Table 1):

Table 1. UNESCO World Heritage.

Category	World heritage in Malaysia
Natural Heritage Sites	a. Gunung Mulu National Park, Sarawak b. Kinabalu National Park, Sabah
Cultural Heritage Sites	a. Melaka and George Town, Historic Cities of Straits of Malacca b. Archaeological Heritage of the Lenggong Valley, Perak
Memory of the World	a.Correspondence of the late Sultan of Kedah (18 1943) b. Sejarah Melayu (The Malay Annals) c. Hikayat Hang Tuah d. Batu Bersurat Terengganu (Inscribed Stone Terengganu)
Global Geopark	Langkawi UNESCO Global Geopark, Kedah
Intangible Cultural Heritage of Humanity	Mak Yong theatre

At the national level, the National Heritage Act 2005 (Act 645) was tabled and endorsed in order to ensure conservation and preservation of the National Heritage; including sites, tangible and intangible cultural heritage, underwater cultural heritage, artefacts and related items. Some of the examples of natural heritage items (apart from those recognized by UNESCO) are:

- Sites (building): St Paul's ruin, Melaka, Carcosa Seri Negara, Sultan Abdul Samad (Supreme Court), Tugu Negara (the National Monument)
- Sites (archealogy): Gua Badak Archeological Site (Cave Drawing), Teluk Kelawar Cave
- Sites (natural): Taman Negara (National Forest - Peninsular Malaysia), Timbalai Triangulation Station
- Objects: Perak Man, Tengkolok Diraja (Royal headgear), Perahu Kemajuan (boat)
- Intangible (dances/performances) : Zapin, Gamelan, Wayang Kulit
- Intangible (traditional games): Wau, Gasing, Congkak
- Intangible (local delicacies): Nasi Lemak, Nasi Tumpang, Nasi Dagang, Laksa Johor

Note however, that the full list can be found on from National Heritage Department's website [1]. The National Heritage Department is the main body governing the regulations and enforcement of the National Heritage Act, besides conducting activities and seminars to create and strengthens awareness in the local community.

3. Efforts in Preservation of Cultural Heritage in Malaysia

Based on survey of the literatures, we present efforts done by various researchers related to cultural heritage preservation in Malaysia:

Malaysia is a member of ASEMUS (Asia-Europe Museum Network); which is a museum network collaboration targeting the use and sharing of objects in museum collection. As part of ASEMUS, there is a section called Virtual Collection of Asian Masterpieces (VCM) with the purpose of sharing museum collections across Asian and European museums. Visitors can search shared museum collections to view images and browse information related to the viewed object [2]. Malaysia participates with the virtual Islamic Arts Museum Malaysia, complete with some of the virtual exhibits as shown as Figure 1.

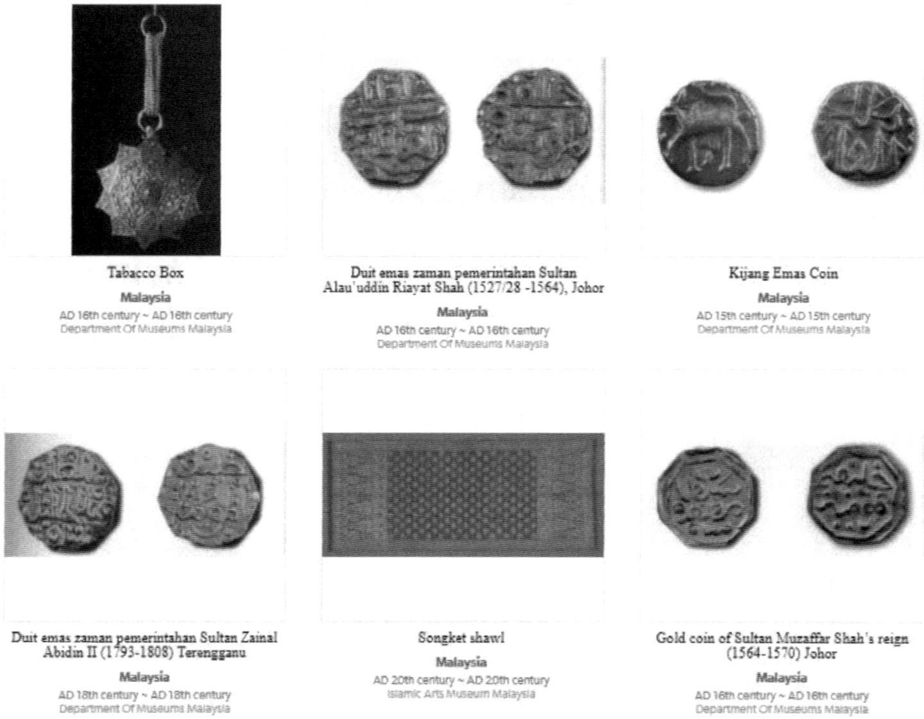

Figure 1. Some of the VCM exhibits from Malaysia [2].

Survey among local people and site observation related to identity of the historic sites (e.g. case study on Kuala Dungun and Taiping) [3] was conducted and another one on Malaysian public's perception on heritage buildings conservation [4]. There are also studies on issues in conservation and redevelopment in areas surrounding historic areas[5], on heritage building conservation (with pilot study)[6], formulation of best maintenance practice guidelines[7], sustainable best practice criteria[8] for Malaysian heritage building conservation, the challenges in conservation practices in Malaysia[9], and the use of Geographic Information System (GIS) technology as digital archive to manage heritage buildings and monuments in Melaka[10]. Similarly, proper asset management on virtual model of the sheltered pavilion of Pak Badol, which has been relocated to the Kelantan Museum of Royal Traditions and Customs for conservation [11] was proposed.

A virtual reconstruction of iconic heritage buildings [12], digital preservation of Malay architectural heritage (Rumah Tok Su – a traditional Malay house in Kedah) [13], recreating the 3D model of traditional Malay house (Teratak Zaaba) [14], virtual preservation of panoramic Kota Kuala Kedah[15] and 3D preservation of the A Famosa Fortress in Melaka[16] are among efforts done in using computer graphics and media technology in preservation of cultural heritage. Apart from that, a study on the potentials and challenges in digitising facial expressions for preservation of a Malay folkdance called Mak Yong [17] was carried out recently.

Studies on performance and establishment measurements on the Geopark concept based on the indicator proposed by Global Geopark Network (GGN)[18] and into the practices of responsible tourism in the Kinabalu Park (UNESCO National Park)[19] were done related to natural heritage sites. The willingness of the tourism to pay the preservation works was also studied and creating value for sustainable tourism in the heritage site [20]. An improved technique for Land Use/Land Cover (LULC) map was proposed for the purposes of environmental monitoring and natural resources management purpose towards conservation of UNESCO Global Geopark - the Kilim Karst Geoforest Park (KKGP) in Langkawi [21].

Among those related to intangible cultural heritage are a study on the visual styles of the Wayang Kulit Kelantan and its capturing methods [22], design and development of interactive virtual Wayang Kulit [23], preservation of Wayang Kulit using multimedia technology[24] into an interactive game[25] and emulating the visuals of Wayang Kulit with Computer Generated Imaginary[26]. Other than that, studies on the determinants of food heritage through food identity [27] and based on age of public perceptions [28] were also done to preserve local food heritage and identity.

Computer game is also used as an indirect way to preserve cultural heritage and create interest and awareness among public. Example of this approach are interactive Wayang Kulit[25], a virtual heritage game called M-Heritage Hunt based on surroundings of George Town, Penang in Malaysia[29], presentation of the history of A Famosa using Game Based Learning methodology[30] and digital Congkak[31]. An AR application was developed to expose users on the masks of the Orang Asli Malaysia (part of the exhibition in the The Museum of Asian Art) as a fun, learning tool [32].

4. Computer Graphics and Media Research for Preservation of Cultural Heritage

We will discuss this topic mainly because our team consists of computer graphics and digital media researchers. Based on previous discussion, there are quite a number of research utilizing computer graphics and multimedia tools and approaches as outlined in the previous sections. It was stated that the first use of graphical representation in archaeology was in the late 1950s [33] using vector-based display. Computer graphics and multimedia technology has progressed to the state of being able to offer digital representation of world cultural heritage. However, it would be costly to set up an online world heritage site; for example, results from a committee appointed by the European Commission estimated 100 billion Euros cost for digitizing European cultural heritage [34].

Computer graphics and multimedia technologies have been used widely in documenting cultural heritage, especially the ones that fall under tangible heritage category. It can be used for the purpose of documentation, reconstruction, visualization, gamification and training towards cultural preservation and awareness. Research onto utilizing computer graphics and media technology into intangible cultural heritage is emerging rapidly.

Based on literature review of cultural heritage preservation in Malaysia, previous efforts were mostly channeled towards preservation of tangible heritage. Digital reconstruction of heritage site/building as a form of tangible heritage preservation is more distinct compared to digital preservation of intangible cultural heritage (ICH). Due to the diversity and subjectivity of ICH, digital preservation of ICH faces enormous technical challenges.

We took up the challenge of venturing into digital preservation of ICH especially for local context. Over the years, there are various research and development projects carried out by our research members that are related to traditional games such as natural interaction using Leap Motion for marble game[35] and Congkak[36] – these are some of the research examples combining affordable hardware (Leap Motion) to cater natural user interaction that incorporate ICH-related contents. Some of the scenes captured from the 3D experience are included as Figure 2 and 3.

Figure 2. Scenes from Marble Game[35].

Figure 3. 3D Congkak[36] incorporating natural user interactions.

An enhancement of virtual reconstruction of historical sites involving the use of augmented reality (AR) technology was done for ancient Malacca [37] – an early settlement in the Straits of Malacca. The use of AR in preservation efforts requires more research such as in object recognition/detection and design of interactions, apart from integration with the 3D environment.

We also work closely with Johor Heritage Foundation in exploring the possibility of using current technology in digital heritage preservation. Based on initial collaboration, we managed to get the involvements of UTM ViCubeLab researchers, undergraduate and postgraduate students to work together with the Johor Heritage Foundation. Experts in traditional dance took part in a series of motion capture data sessions (using marker (Figure 4) and markerless multi-camera systems (Figure 5)) and video recording (Figure 6) to get reliable Zapin dance data for further use.

The captured Zapin dance data were used in various research, for example a study on suitable algorithms to generate a more natural facial expressions for motion capture data re-use in 3D animation[38], to propose a method of extracting the 3D dance motion data of Zapin traditional dance from video data using keyframe animation extraction method[39] and a framework for constructing a 3D motion and skeleton using a monocular video source (2D video) targeted for traditional dance motion data extraction[40].

IOP Conf. Series: Materials Science and Engineering **979** (2020) 012008 doi:10.1088/1757-899X/979/1/012008

Figure 4. Expert involvement in a marker-based motion capture session.

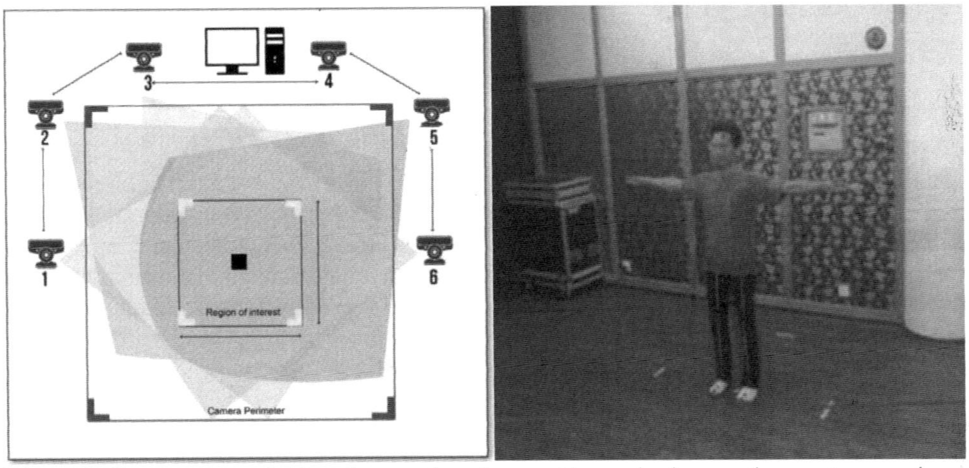

Figure 5. Camera setup (left) and expert involvement in markerless motion capture sessions.

Figure 6. Expert involvement (left) in video recording sessions.

On the lighter side, we managed to integrate heritage elements into our final year projects. Some of the examples are the study of fabrics and motives used for dance costumes as shown in Figure 7 (this was based on the visit to textile gallery at Johor Heritage Foundation) along with motion capture data re-use for animation [41], photogrammetry for artifacts reconstruction [42] and virtual museum on handheld device [43].

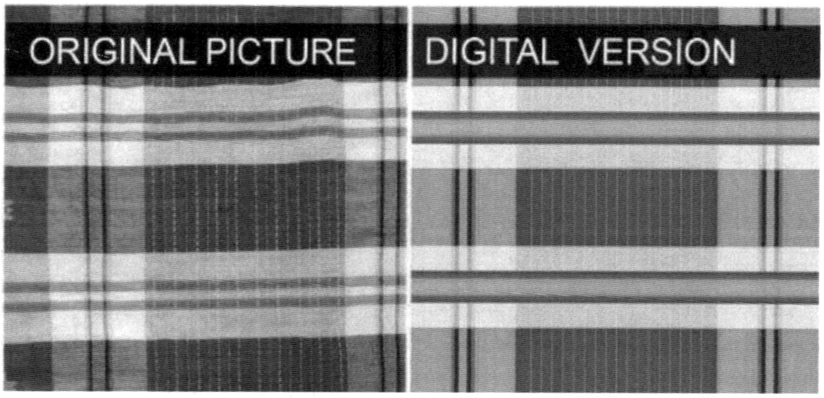

Figure 7. Original motive on textile (left) and the digital adaptation (right) [41].

5. Conclusion and Future Work

As a conclusion, we have seen the diverse cultural heritage available in Malaysia. Based on what were found from the literature, most of the preservation efforts (including usage of computer graphics and media) reported were mostly centered around tangible cultural heritage. Since computer graphics and media technology has so much to offer, there are endless possibilities for cultural preservation efforts digitally both for tangible and intangible cultural heritage preservation. Our team is currently working with Southeast Asian and European partners towards intangible cultural heritage preservation using computer graphics and computational science methods.

Acknowledgement

The authors would like to express our appreciation to Universiti Teknologi Malaysia (UTM) and Ministry of Education (MOE) for financial support and the opportunity to carry out this research. This project was supported by Research University Grant - UTM ER [Vot Number: 19J10] initiated by UTM. Special thanks to the Johor Heritage Foundation for the direct involvements and to all other parties that either directly or indirectly contributed to the success of this research.

References

[1] Department, N.H. *Jabatan Warisan Negara.* [cited 2020 1 Oct 2020]; Available from: www.heritage.gov.my/.

[2] (ASEMUS), A.-E.M.N. *The Virtual Collection of Asian Masterpieces (VCM).* 2013 [cited 2020 23 October 2020]; Available from: http://masterpieces.asemus.museum/museum/list.nhn.

[3] Lai, L.Y., I. Said, and A. Kubota, *The Roles of Cultural Spaces in Malaysia's Historic Towns: The Case of Kuala Dungun and Taiping.* Procedia - Social and Behavioral Sciences, 2013. **85**: p. 602-625.

[4] Azhari, N.F.N. and E. Mohamed, *Public Perception: Heritage Building Conservation in Kuala Lumpur.* Procedia - Social and Behavioral Sciences, 2012. 50: p. 271-279.

[5] Said, S.Y., H. Aksah, and E.D. Ismail, *Heritage Conservation and Regeneration of Historic Areas*

in Malaysia. Procedia - Social and Behavioral Sciences, 2013. 105: p. 418-428.
[6] Azizi, N.Z.M., et al., *Recurring Issues in Historic Building Conservation.* Procedia - Social and Behavioral Sciences, 2016. 222: p. 587-595.
[7] Mohd-Isa, A.F., Z. Zainal-Abidin, and A.E. Hashim, *Built Heritage Maintenance: A Malaysian Perspectives.* Procedia Engineering, 2011. 20: p. 213-221.
[8] Sodangi, M., et al., *Best Practice Criteria for Sustainable Maintenance Management of Heritage Buildings in Malaysia.* Procedia Engineering, 2014. **77**: p. 11-19.
[9] Azizan, M.A., et al., *The challenges in conservation practices in Malaysia: A study in UNESCO heritage site, Georgetown, Penang, Malaysia.* AIP Conference Proceedings, 2020. 2213(1): p. 020277.
[10] Musa, S.H., et al., *The Development of GIS-Based Digital Archive for Heritage Building in Melaka World Heritage Site (UNESCO), Malaysia.* International Journal of Engineering Advanced Research, 2020. 2(3): p. 1-16.
[11] Idris, W.M.R.W., et al., *Modelling Virtual Heritage Environments Using Hierarchy-Embedded Virtual Objects.* International Journal of Multimedia and Ubiquitous Engineering, 2015. 10(4): p. 259-272.
[12] Esmaeili, H., P.C. Woods, and H. Thwaites, *Virtual Preservation of Contemporary Architectural Heritage in Developing Countries in Absence of Protection: Digital Reconstrction, Recording, and Archiving before Complete Disappearance*, in *Proceedings of the 2015 11th International Conference on Signal-Image Technology & Internet-Based Systems (SITIS)*. 2015, IEEE Computer Society. p. 796-803.
[13] Ibrahim, N. and K.A. Azmi. *Preserving Malay architectural heritage through virtual reconstruction.* in *Digital Heritage International Congress (DigitalHeritage), 2013*. 2013.
[14] Ibrahim, N., et al. *Cultural Heritage Preservation: 3D Modeling of Traditional Malay House using Hidden Surface Removal Approach.* in *Proceeding of International Conference on Software Engineering and Computer Systems, ICSECS*. 2009.
[15] Jabbar, E.A. and M.J. Jusof. *Virtual heritage interpretation through animated panoramic illustration of Kota Kuala Kedah.* in *Virtual Systems & Multimedia (VSMM), 2014 International Conference on*. 2014.
[16] Izani, M., et al., *Digital preservation of a famosa fortress in malaysia*, in *Proceedings of the second workshop on eHeritage and digital art preservation.* 2010, ACM: Firenze, Italy. p. 7-12.
[17] Idris, M.Z. and N. Musa. *Beyond Digitalisation: Facial Motion Capture for Mak Yong through the Perspective of Aesthetic Experience and Uncanny Valley*. 2020.
[18] Fauzi, N.S.M. and A. Misni, *Geoheritage Conservation: Indicators Affecting the Condition and Sustainability of Geopark – A Conceptual Review.* Procedia - Social and Behavioral Sciences, 2016. 222: p. 676-684.
[19] Tay, K.X., et al., *Comprehending the responsible tourism practices through principles of sustainability: A case of Kinabalu Park.* Tourism Management Perspectives, 2016. 18: p. 34-41.
[20] Kamaludin, M., N.F. Haron, and M.A.S. Razali, *Tourists' Preferences for Preservation of World Heritage Site Stadthuys, Malacca.* Journal of Environmental Management and Tourism, 2020(2):11 p. 281-289.
[21] Halim, M.K.A., et al., *Land use/land cover mapping for conservation of UNESCO Global Geopark using object and pixel-based approaches.* IOP Conference Series: Earth and Environmental Science, 2018. 169: p. 012075.
[22] Khor, K.-K. and Y.M. Chan. *A Study on the Visual Styles of Wayang Kulit Kelantan and Its Capturing Methods*. in *Computer Graphics, Imaging and Visualization, 2009. CGIV '09. Sixth International Conference on*. 2009.
[23] Talib, A.Z., et al., *Design and Development of an Interactive Virtual Shadow Puppet Play*, in *Arts*

and Technology: Second International Conference, ArtsIT 2011, Esbjerg, Denmark, December 10-11, 2011, Revised Selected Papers, A.L. Brooks, Editor. 2012, Springer Berlin Heidelberg: Berlin, Heidelberg. p. 118-126.

[24] Dahlan Abdul, G., *Preserving Wayang Kulit for Future Generations*, I. Mohd Sidin Bin Ahmad, Editor. 2011. p. 70-74.

[25] Ghani, D.B.A., *Wayang fighters: preserving the art of wayang kulit (shadow puppets) through game application*, in *Proceedings of the 9th International Conference on Ubiquitous Information Management and Communication*. 2015, ACM: Bali, Indonesia. p. 1-7.

[26] Khor, K.-K., *A Methodological Framework to Emulate the Visual of Malaysian Shadow Play With Computer-Generated Imagery*, in *Innovative Computing, Optimization and Its Applications: Modelling and Simulations*, I. Zelinka, et al., Editors. 2018, Springer International Publishing: Cham. p. 321-336.

[27] Md Ramli, A., et al., *Determinants of Food Heritage towards Food Identity*. Environment-Behaviour Proceedings Journal, 2016. 1(1): p. 207-216.

[28] Md Ramli, A., D. Khairunisa, and M. Salehuddin, *The determinants of food heritage based on age in Klang Valley*. 2020. 12: p. 396-411.

[29] Kiam Lam, T., L. Chen Kim, and T. Abdullah Zawawi bin Haji, *Mobile Virtual Heritage Exploration with Heritage Hunt with a Case Study of George Town, Penang, Malaysia*, in *Modern Entrepreneurship and E-Business Innovations*, N. Süphan, Editor. 2013, IGI Global: Hershey, PA, USA. p. 115-126.

[30] Izani, M., M. Grant, and A. Razak, *A proposal for mobile game based learning to promote knowledge of the a famosa fortress*, in *Proceedings of the 4th international conference on Progress in Cultural Heritage Preservation*. 2012, Springer-Verlag: Limassol, Cyprus. p. 720-728.

[31] Mohamad, A.F.H., J.J.M. Supian, and M.S.M. Ribuan, *Digital Congkak: The Art, Play and Experience Framework*, in *Proceedings of the 10th International Conference on Ubiquitous Information Management and Communication*. 2016, ACM: Danang, Viet Nam. p. 1-6.

[32] Sidyawati, L., et al., *Nusantara Mask Heritage Malaysia: Infographic Application Development of Masks of Malaysian Indigenous Tribes at the Museum of Asian Art Malaysia based on Augmented Reality as Media of Tourism Education*. Jurnal IPTA (Industri Perjalanan Wisata), 2019(2): p. 163-173.

[33] Cowgill, G.L., *Computer applications in archaeology*. Computers and the Humanities, 1967. **2**(1): p. 17-23.

[34] Arnold, D., *Computer Graphics and Cultural Heritage: From One-Way Inspiration to Symbiosis, Part 1*. IEEE Computer Graphics and Applications, 2014. 34(3): p. 76-86.

[35] Masaan, M.F.M. and N.M. Suaib. *Marbles Game in 3D Virtual Environment using Hand Gesture Interaction*. in *INNOVATIONS IN COMPUTING TECHNOLOGY AND APPLICATIONS*. 2018. School of Computing, UTM.

[36] Zolkifly, M.H.I.M. and N.M. Suaib. *Hand Gesture Integration of Congkak in 3D Virtual Environment*. in *INNOVATIONS IN COMPUTING TECHNOLOGY AND APPLICATIONS*. 2018. School of Computing, UTM.

[37] Ismail, A.W., *User Interaction Technique With 3D Object Manipulation in Augmented Reality Environment*. 2011, Universiti Teknologi Malaysia.

[38] Ahmad, M.A., N. Mohd Suaib, and A.H. Ahmad Rahman, *Position-Aware 3D Facial Expression Mapping Using Ray Casting and Blendshape*, in *Encyclopedia of Computer Graphics and Games*, N. Lee, Editor. 2020, Springer International Publishing: Cham. p. 1-4.

[39] Albakri, I.F., et al. *3D Keyframe Motion Extraction from Zapin Traditional Dance Videos*. 2020. Singapore: Springer Singapore.

[40] Azmi, N.M.W., et al. *3D Motion and Skeleton Construction from Monocular Video*. 2020. Singapore:

Springer Singapore.

[41] Mohd Faudzi, M.S., *Motion Capture Data Reuse for Zapin Melayu Johor Preservation for Children*, in *School of Computing*. 2019, Universiti Teknologi Malaysia: UTM Skudai.

[42] Engitaravanan, L.R., *3D Modelling using Photogrammetry for Cultural Heritage Preservation*, in *School of Computing*. 2020, Universiti Teknologi Malaysia: UTM Skudai.

[43] Zainuddin, S.N.L., *Interactive Virtual Museum Using Head Mounted Device*, in *School of Computing*. 2020, Universiti Teknologi Malaysia: UTM Skudai.

IOP Conf. Series: Materials Science and Engineering **979** (2020) 012009 doi:10.1088/1757-899X/979/1/012009

Finger-Ray Interaction using Real Hand in Handheld Augmented Reality Interface

C S Yusof [1,2] *, N A A Halim[1,2], M N A Nor'a[1,2] and A W Ismail[1,2]

[1] Mixed and Virtual Environment Research Lab (mivielab), ViCubeLab, Universiti Teknologi Malaysia, 81310 Johor, Malaysia
[2] School of Computing, Faculty of Engineering, Universiti Teknologi Malaysia, 81310 Johor, Malaysia

* suhaimi@utm.my

Abstract. Augmented Reality (AR) is the technology that augments the real world by using virtual information of 3D objects overlaid on a view of the real environment. This will create an immersive and intuitive experience. As handheld devices are widely used and now has climb with increasing demand of specifications, its interaction based on touch is a natural and appealing input style for the application of AR. Furthermore, most heuristic studies of interaction in AR usually focus on interactions with AR Target close to the user, generally within arm's reach. As the user's move farther away, the effectiveness and usability of the interaction modalities may be different. This study explores handheld AR interaction using real hand gesture at a distance in a room-scale setup. Our aim is to investigate the effectiveness of performing selection while being far away from the 3D object and performing selection task towards object that is being occluded. As object manipulation is one of the key features to explore for mobile handheld AR, we hope our study can give some contribution towards its practical use, especially for real-world assembly structure. Hence, our paper proposes finger ray interaction technique for real hand in handheld AR interface to work for selecting objects at distances ranging from 3 feet to 8 feet and when the object being occluded ranging from 20% to 80% occluded.

1. Introduction

In this advance technology era, Augmented Reality (AR) application in mobile devices has bloomed, making it widely spread and is used for a growing range of application. The increasing performance of computational, cameras, processors, displays, and various sensors in mobile devices, making it possible for researchers to discover the full capabilities of handheld AR. The transformation of learning caused by AR provides exciting opportunities to build learning environments that are engaging, realistic and fun [1]. For researchers and developers, the rapid development of handhelds has opened up plenty of scope to explore new and previously unattainable ways to provide mobile AR experiences [2].

Currently, people enjoy using AR because it helps to give them a better understanding of some of the information rather than just by looking at the plain view of the information. For example, the application from ARTutor [3], enables students to verbally ask questions and obtain responses based on the book's content. This implies that the framework is sufficient for distance learning and independent learning, thus encouraging students to learn and understand the subjects better.

Meanwhile, Heb@AR [4], which is not yet covered by the literature, explores the use of handheld

AR as an additional method for technical education in academic midwifery education. Another example, the usage of VR and AR in the dentistry not only use the system for learning and teaching from their perspective but also training their skill and improve the hand-eye coordinate indirectly help the user to correct the posture and skill [11].

However, there are general problem in handheld device, such as limited screen size, and one hand interaction (in which the other hand need to hold the device). On smartphones and tablets, nearly all handheld mobile AR systems are introduced. This is complicated by the fact that the tracker (the part used to monitor the AR marker), which is a built-in camera, [24] often used on the same device (smartphone or tablet) as the rendering systems linked to the display screen.

As part of a larger body of work by conducting on mobile AR authoring, Sambrooks et al. examine the impact these various device sizes have on user interaction with AR content [2]. Their experiment is to compare AR interaction for three different handheld devices which were chosen as prototypical representations of three categories: smartphone, mini tablet, and tablet. In particular, the screen size was noted as offering a clearer view of the on-screen crosshair and virtual objects. Meanwhile, in the finding of Hürst et al., some users complained that doing the motions with one hand and clicking on the touch screen when holding the computer with the other hand was uncomfortable and that it put an extra cognitive burden on them [24]. In handheld AR, pointing techniques are impaired by the instability of the augmented scene that the users interact with on screen [9].

AR head-mounted displays (AR-HMDs), such as the Microsoft HoloLens, support immersive applications which embed interactive virtual content throughout the visible environment. Nonetheless, in recent years, many commercially usable AR HMDs have like Microsoft HoloLens, has come to market including the Microsoft HoloLens, Daqri Smart Glasses, Meta 2, and Magic Leap [4-7]. For these applications to give smooth, seamless interactive experiences, creators must craft intuitive and efficient interactions with the virtual content at extended spatial scales, including virtual objects to be well beyond a user's reach [7].

In the real world, AR apps also use virtual content to augment objects, so this decontextualization can generate cognitive challenges. Instead, to test accuracy and ease of use in communicating with distant objects in AR, we concentrate on selecting and manipulating material that remains at a fixed distance and building on previous findings in these spaces [10]. Through the use of head-mounted displays and monitoring devices, such as data goggles, and on desktop VR settings with a keyboard and mouse, three-dimensional interaction techniques have been extensively studied in interactive virtual environments. Several researchers have studied 3D interaction techniques that, particularly for desktop and large-scale interactions, approach the richness of reality. The benefit of controlling 3D objects using mobile devices in augmented reality scenarios is the ability to use physical gestures for interaction. This technique is a normal way of placing objects in the scene, as it mimics the actual behavior [14].

The field of interaction and display is restricted due to the limited size of the display. During interaction, users often cover the target with their fingertips when using multi-touch input [16]. The size of the target on the limited screen that user can touch is really small. This constrain is becoming a crucial problem with the growing of complexity of AR scenes. Another case is that when the targets are obscured from each other in a 3D AR environment due to the variation of its spatial depth, the selection by touch becoming hard to be perform. Depth is also the main issue when we do interaction using 2D interface for 3d environment. Thus, an accurate of pixel-level selection in occlusive or compact areas with various sizes of the targets making it difficult to perform [17].

For these purposes, although they are partially or entirely obscured by other virtual targets, the selection performed must include the possibility to disambiguate and able to select desired targets with preciseness. With current selection techniques, choosing a target can be hard or even difficult with targets of strong visual resemblance in compact virtual worlds. It should retain some spatial information which will enable disambiguation. It also includes the interactions of a single touch to ease the user interaction and eliminate faults conveyed by incorrect feedback of touch input in order to have a unique selection of a target. We use ray-casting selection technique for the handheld AR interface to resolve these interaction issues, although the target is in small sizes or being occluded, it can precisely select targets.

2. Related Works

In 3D user interfaces (3DUIs), selection is one of the universal interaction tasks [12] and has been extensively studied by researchers and developers. While several guidelines have been created for developers of 3DUIs, the research has clearly shown that there is no best selection technique for all situations. Natural and intuitive interaction techniques such a spatial gesture input are required by user instead of simple pointing and clicking and manipulating 3d content from a 2d interface to have better experience with AR [18]. Manipulating 3D objects is a complicated task that requires multiple degree-of - freedom (DoF) control for object selection, translation, rotation and scaling in AR environments. In addition, virtual objects are intangible, and only by whole-body observation or moderate by a mobile device can contact with them be accomplished [12].

Metehan et. al. [20] presented a novel concept for AR in cultural heritage sites: Distant Augmented Reality in which they explore a new possibility; proposing drones as a medium to communicate through AR. Based on the relevant task criteria (i.e., distance and size of the target) and the layout of the environment, such as occlusions of the target and the density of the scene, the efficiency and usability of a selection technique vary greatly, as seen in the literature. It often consists of two sub-tasks: determining the goal and the verifying step of the optional collection. For virtual environments (VEs), as there are several selection methods that have been developed, we set apart the related works into selection with handheld mobile touch interface (device based) developed for VEs that can be modify to AR, and with mid-air gesture-based interaction technique selection. The fact that almost all handheld mobile AR systems are implemented on smartphones and tablets by means of the tracker (used to monitor the AR marker) is the built-in camera that has also been used as the rendering device associated with the display screen, all together on the same platform, complicated the implementation of 3D object manipulation of virtual objects in handheld mobile AR.

Ray-Casting [13] is a frequently utilized technique of pointing-based where a virtual ray projecting from the input device or real hand when user points to target in the scene. For Ray-Casting, this is a bigger problem than for other methods because at the tip of the rays, small hand motions are amplified allowing Ray-Casting to be less accurate as targets get farther from the consumer. These issues make Ray- Casting difficult to use when the targets have a small visual size [14], as selecting such targets by pointing requires high levels of precision. In order to address these issues, a number of improvements have been proposed. Even though such techniques improve selection performance in general, they can have a negative effect in very cluttered environments. Cone-Casting [15], for example, extends ray casting by adding a cone-shaped volume to the ray to make it easier to select distant targets.

Current 3D target selection approaches in AR for mobile generally use the basic specifically pointing metaphor activated on the mobile screen by a single contact case. [16]. However, these methods lack accuracy due to the finger size of users in a cluttered mobile world. Selection techniques of 2D that able to control the issue of finger being occluded on the screen have been confer to improve accuracy in 2D touch interface [19]. Dual-Finger Offset and Dual-Finger Midpoint. Both of these needed the precision selection of two hands and resulting of it's to be not suitable to be applied in a scenario of mobile AR where there is just one hand accessible for input. Two selection techniques specifically outline were proposed in reference for handheld AR [19], using two fingers to pick the targets that is small and partially occluded in dense AR. The procedure is encouraging but cannot be extended to the extreme screen corners to select targets. As this reduces the already small space for interaction, this is not an ideal solution for a handheld device, for example smartphone. In addition, in [19], by differentiating the selection point from the physical touchpoint, the selection in dense mobile AR using a single touch was explained by the authors. Thus, the occlusion of the target by the finger can be evaded, but statically calculated for the offset.

3. Implementation

Realism [15] is a crucial factor in determining the right measurement of distance and perception of depth. There are significant variables that affect efficiency and ease of use when developing a selection system for handheld AR.

On these displays, manipulation directly suffers from minimal preciseness, issues of being occluded, and restrictions of the scene element's size due to the physical constraints of the mobile device size and the constraints posed by human fingers. We discuss the criteria for accurate selection of one-handed, handheld AR setting in the compact environment for this section. Then we will use the ray casting technique to compare the interaction between the handheld mobile app and the actual hand gesture. This is to equate the techniques' efficiency and usability with the increasing in distance, rendering the target 3D object smaller. Mid-air free-hand gesture interaction has been considered as a promising input method for MAR systems to solve the above problem while providing a richer interaction experience [23].

3.1. *Requirement and User Setting*

Our aim is to achieve selection that is precise for handheld AR in dense environment with one-handed, even though the distance makes the 3D object smaller. We summarize the requirements as follows. A built-in camera of the handheld device is used as display technique. This camera is used for video streaming the target marker. A few configurations have to be made before it can be used. If the target marker is recognized as a registered marker, the AR 3D object will appear at the center of the marker on the display screen of the unit. In order to run the application on the handheld device without any problems or errors, the right API is required. We use Leap Motion, a depth sensor device to track the gesture input.

The Leap Motion Controller is a device for optical hand tracking that captures with unprecedented precision the movements of your hands. During recognition, Leap Motion enables the process to read depth data for the application. Then orientations and positions were generated by the computer. The technique for tracking presented is focused on functionality. Figure 1 display a natural feature tracking (NFT) process, which also the feature-based tracking technique. The tracking technique includes the registration in the real world of the virtual object over the actual marker.

Figure 1. Standard marker for NFT.

Compared to immersive VEs, independent monitoring of the interaction device from the user is not obtainable with the head. Thus, a single device combines input and output devices. The gesture interaction is analyzed at this point to engage between the user with the virtual space of AR. A form of interaction using hand motion can be done to engage with the virtual object naturally. The user will use their finger to touch the virtual elements directly. By pointing their finger towards the 3D object that is visible on the marker, users may perform a gestural interaction. In the project, therefore, as a metaphor, the method of hand gesture recognition for feedback are necessary for the engagement of the user with the virtual object in AR space.

3.2. *Leap Motion Gesture Tracking*

As a tracking device, Leap Motion enable the process to read depth data from the application, thus the user hand location can be detected in the actual world and mapped to the virtual world [22]. Below in Figure 2, it displays the gesture tracking standard flow of Leap Motion, which enables to read depth data during recognition of the application. Then, positions and orientations were generated. It observed the location of the user's hand in the actual world and maps it to the virtual world. The modelling method is needed to show a virtual hand skeleton and to allow interaction signals, and the rigid body was applied to the virtual hands 3D model. After this phase has been completed, both the gesture input and the dynamic gestures will be produced.

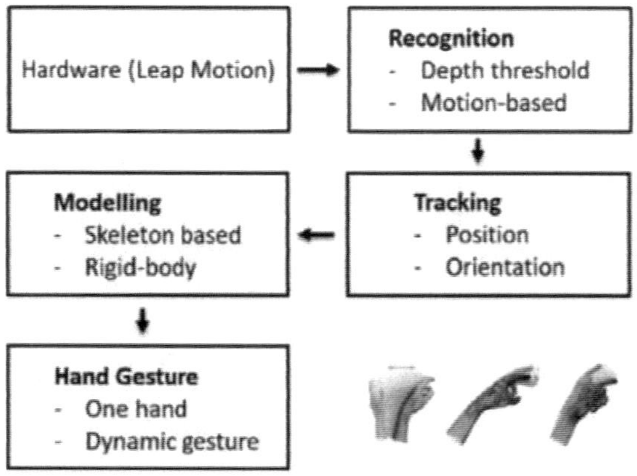

Figure 2. Gesture tracking standard flow [22].

Because of the wide space occupies on the display due to the user's fingertip, finger's input of touching can be imprecise. Since only one hand is required for contact, it is not possible to apply complex multi-hand movements to increase the preciseness of selection.

Figure 3 below shows the system workflow of our current prototype in details. One of the problem that we face, is regarding the data calibration between build in camera of the handheld device and the Leap Motion itself. To avoid this issues and have better result of finger tracking, we attach the Leap Motion behind the mobile device and near the build-in camera of the handheld device. Leap Motion, a hand tracking device for Virtual Reality (VR), will provide us the information of hand tracking which the most important information is the depth. Leap Motion is used in our interaction system for marker-less Augmented Reality to detect the hand.

By using Leap Motion, 3D hand location, gesture and direction can be obtained and transferred to the Unity 3d game engine. We are using Photon Unity Networking (PUN) library to transmit the finger tracking data between devices. For better understanding we divide the process into 4 phase which are 1) Master Client will create a session to receive hand tracking information from the Leap Motion, 2) Using photon library, the data of hand tracking collected from Leap Motion will then be transmitted to the Handheld Client(mobile device) application 3) User now can view the finger movement from both Master Client and Handheld Client 4) User now can do the selection process by selecting the item that appear on top of the AR Target Maker.

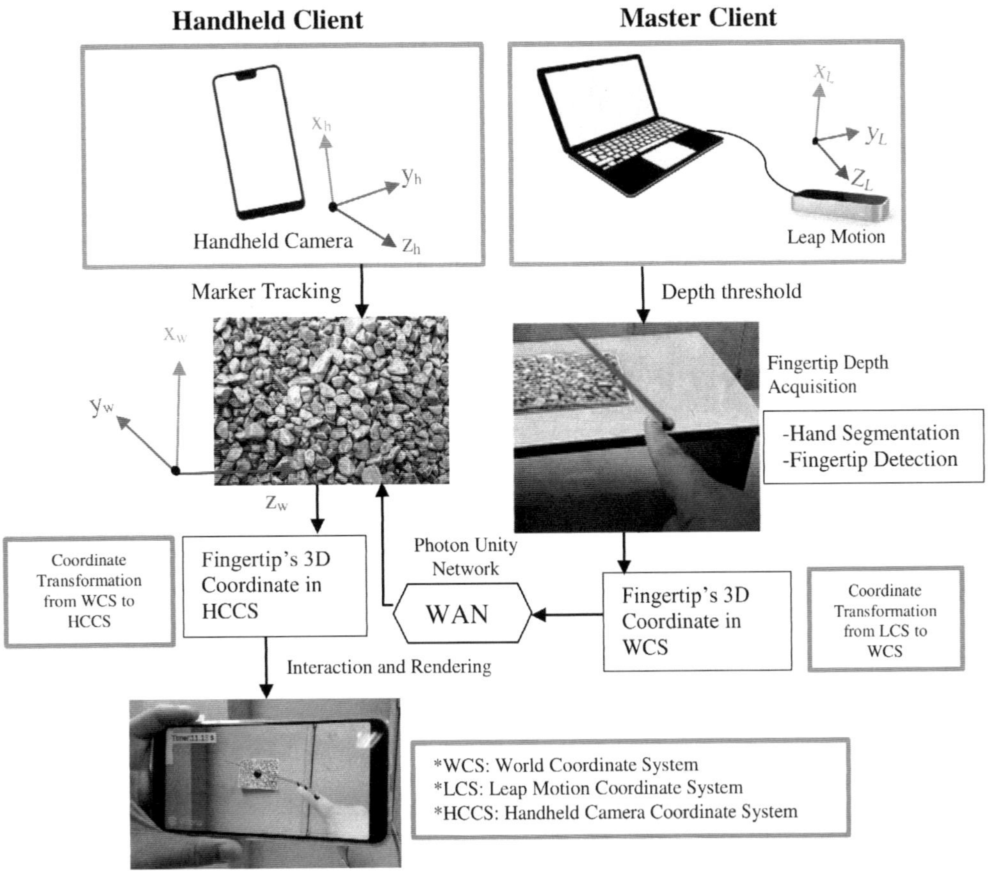

Figure 3. System workflow in details.

3.3. *Proposed Selection Method*

As stated previously, we come to develop the below selection methods in enabling the selection to be perform precisely although in handheld AR with dense environment of one-handed.

1) As in dense virtual scenes the targets might be invisible or partly occluded, it is crucial to have a technique that hold up the process of selecting these target. Moreover, targets can be extremely indistinguishable in figurative display. Therefore, multiple selected targets need to be described as proper spatial sense which encourages disambiguation of the target while considering the minimal display available.
2) In viewing context of mobile use, process of interaction does not presume that supplemental of physical resources (for example, extra devices of 3D pointing) are present for the interaction
3) In comparison to inexplicably (outside-in) approaches to 3D tabletop processes, the egocentric perspective should be based on handheld mobile AR interaction techniques.
4) The interactive methods should provide the consumer with sufficient input, either visually or in another type. For instance, the user should experience constant visual / physical connections during the manipulation. [21].

A finger-ray has been proposed based on these design for selection, where a ray is generated based on the finger-pointing using user's bare hand. The user achieves this method by moving a ray-cast point projected from the central position of the field of view; the finger detected will produce the ray where it was corresponding to the user's finger movement. This enables the user to manipulate a 3D object appeared on the AR marker right away in a similar manner to manipulating a close object. However, this paper needs to execute a test system to the feasibility of our proposed method before implementing the method in the AR application to manipulate the 3D object. The interaction is a simple selection, just to test the ray precisely hit the object based on the distance.

The most popular target pointing technique in VR is ray-casting, and AR has been implemented the VR interaction which introduced to be accessible in augmented environments. The downside is that output is influenced by the precision of the pointing technique and the controller on small and distant targets. Current pointing facilitation techniques are currently only used in the sense of the virtual hand and pointing is done using their bare finger in AR by users.

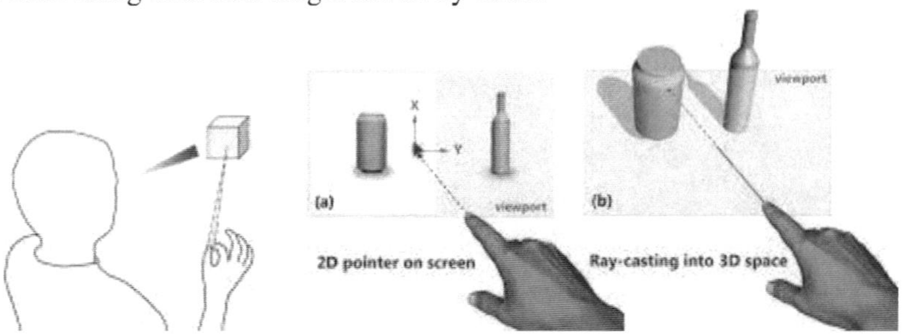

Figure 4. A finger-ray selection method.

Pointing a ray same like a laser pointer behavior. The origin and orientation are defined by those of the input device with 6 degrees of freedom. Figure 4 shows a user uses their finger to hit the cube by projecting the ray to perform a selection. Figure 4 (a) shows a 2D pointer on the screen relative to the x-axis and y-axis for handheld, however Figure 4 (b) shows the ray- casting into the 3D scene.

4. Finger-ray Interaction

In this section, we present the development of our project in progress. Figure 5 exhibited that Ray-Casting is an effortless, method of one step selection and used in VEs (immersive) and 3D computer very actively. In the virtual scene, a virtual ray is cast into; when it crosses them, targets are chosen. Ray-Casting is quick and effective for targets in near-range, however with targets that is small and occlusions at a greater distance, it has issues.

4.1 Selection by distance

We use the following adaptation in using Ray-Casting for a handheld AR environment: with a single touch event, Ray-Casting is provoked on the display; coordinates of the 2D display are projected back into 3D space, and a virtual ray is cast from the location of the virtual camera in the direction of the back-projected 3D point in the handheld AR scene.

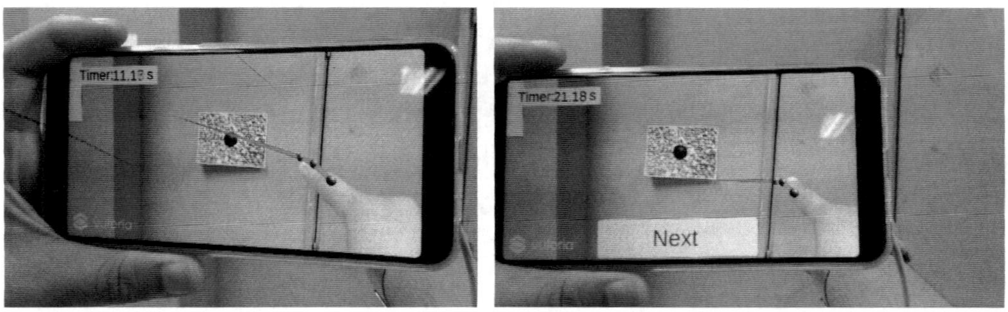

(a) User selecting 3D object using Ray-Casting technique (b) A button will be displayed after the object has

Figure 5 Selection experiment using a finger-ray interaction.

A timer is shown on the top left of the device's screen, providing information on the time taken for the 3D object to be selected by the user. A 'Start' button is displayed on the screen for the user to touch, which will then start the timer. This will give us a suggestion of estimation on time taken for the user to select the object as the distance increases. The camera will then display the index finger skeleton that is being retrieved from the leap motion attach at the back of the handheld device.

To select the desired virtual object, our ray casting technique project a ray-casting pointer from the tip of the index finger skeleton. When the pointer ray-casting hit the object, it will select the object, and the timer will be stopped. When the object has been selected, the user will receive sound feedback, and it will also trigger haptic feedback to notify the user that the object is being selected. This will provide a better AR experience for the user in digesting the information given.

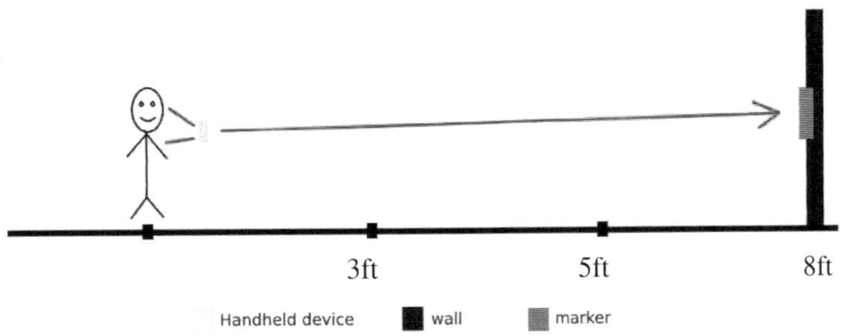

Figure 6. The distance setting for the experiment.

As shown in Figure 6 above, we test the finger-ray interaction with three different distance, which is 3 feet, 5 feet and 8 feet. As the distance decreases, the user's speed is faster in selecting the object as compared to selecting the object at a further distance. As shown in Figure 7, the selection of 3D object with handheld mobile touch interface is more familiar for the user. However, due to the small surface area for the object (the object being smaller as the distance increases), the object selection becomes harder. User needs to touch the mobile screen a few time, before able to select the 3D object.

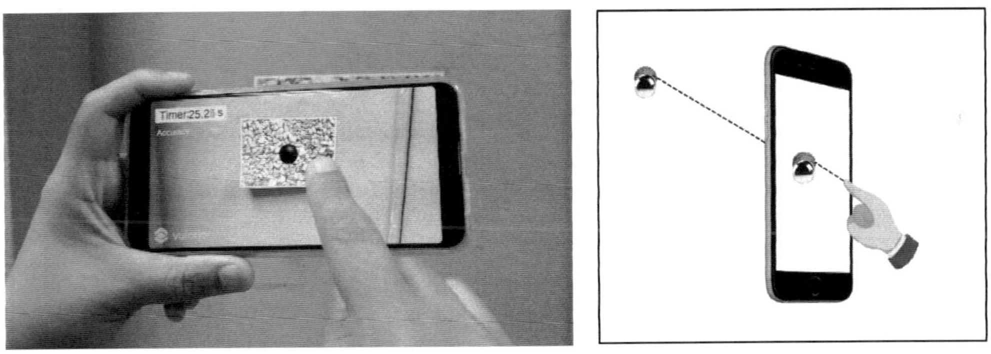

(a) (b)

Figure 7. (a) common touch-based interaction (b) ray casting selection process.

4.2 Selection by occlusion condition

Occlusion reduce the surface of object that will be available to detect. As referred to Figure 7, the user touches the touch screen. Ray-Casting is quick and precise for targets in near-range, but it has issues with small targets and occlusions at a greater distance. Thus in our project we test the finger ray interaction towards the occluded object at 20%, 50% and 80% occlusion as shown in Table 1. For occlusion, user will be standing 3 feet from the marker. As shown in Figure 8 below, probability for user to touch more than once to select the object increases as the percentage of the object being occluded increases.

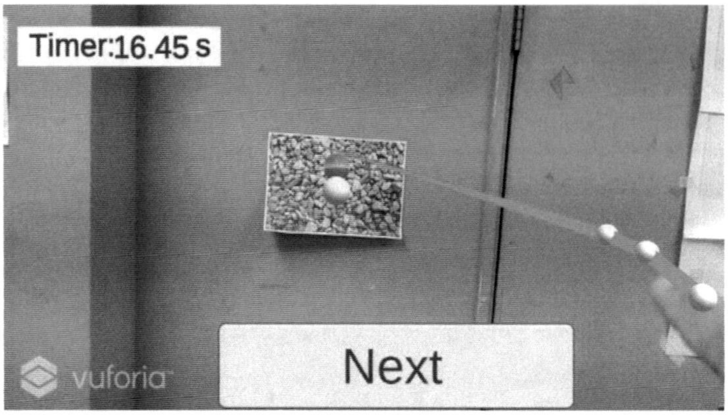

Figure 8. Selecting the occluded object (20% occluded).

Table 1. Occlusion data.

% Occluded	Distance from centre(h)
20	0.8
50	0.5
80	0.2

4.3 Experimental Task

As Yin et al. state in their paper [13], we also plan to invite participants who have experience with AR with three factors:

- distance of the target (Distance 3feet, 5feet, 8feet)
- selection technique of Ray Casting
- occlusion level of the target (20%, 50%, and 80%)

Participant will need to select as accurately as possible for the targets and do it as fast as they can. These variables will be evaluated for 12 participants of every technique. Therefore, in total, the design of the experiment will have resulted in 12 participants × 2 target × 3 condition × 2 techniques × 3 repetitions for each group = 432 total trials. The result would be 36 trials per participant. We will then gather the distance and level of occlusion for each chosen target to be studied. We also will obtain time and error rate for each trial.

Table 2. Post-Test Questionnaire.

#	Questions in Usability Questionnaires
Q1	I was performing well using this technique.
Q2	This interaction technique is easy to use.
Q3	This interaction technique is easy to learn.
Q4	This interaction technique is intuitive.
Q5	This interaction technique is natural.
Q6	This technique is useful for completing task.
Q7	This technique has no mental stress.
Q8	This technique has no physical stress.

A post-questionnaire will be given to participants as the experiment of the technique completed (see Table 2). The 7-point Likert scale will be used to present all the questions, where 1 was the most negative response and 7 was the most positive response. They will be asked to rank the techniques from 1 (most preferred) to 6 (least preferred) after the technical experiments were all completed.

5. Summary

We explore the occlusion and distance issue that may affect the far object in AR, and the interaction was focusing on AR. We compare the efficiency of applying our technique with handheld mobile touch interface. For selecting virtual objects, this is to analyze performance and preference. Both articulated freehand gestures and facilitated remote handheld interactions contributed to interactions that were more powerful than voice control.

People greatly favored embodied interactions, however, and considered them to be faster and easier to use. These results provide initial insight into the nature of distal interaction systems in AR, where interactive environments provide experiences with digital material embedded in the physical world on an environmental scale. Iin the future, we would like to continue and extend the experiment by considering more complex interaction such as translation and rotation, and study the efficiency of the interaction. We hope from this study, a better understanding of the finger-ray interaction technique can be achieved and thus will contribute into the growth of research on the possibilities of effective AR applications for example AR based shopping application, urban planning or any other interesting possibilities.

Acknowledgement

We would like to convey our gratitude to UTMER Grant (Q.J130000.3851.19J22) for the funding. Special thanks to Mixed and Virtual Reality Laboratory (mivielab), ViCubeLab at Universiti Teknologi Malaysia (UTM) for the technical supports and facilities. We are also thankful to the parties that contributed the fundamental knowledge in this paper.

References

[1] Haque, R., Islam, M. M., Salma, S., Al Jubair, M. A., & Weng, N. G. (2020, January). Extracting Relevant Information Using Handheld Augmented Reality. In Proceedings of the International Conference on Computing Advancements (pp. 1-6).

[2] Sambrooks, L., & Wilkinson, B. (2015, January). Handheld Augmented Reality: Does Size Matter?. In Proceedings of the 16th Australasian User Interface Conference (AUIC 2015) (Vol. 27, p. 30).

[3] Blattgerste, J., Luksch, K., Lewa, C., Kunzendorf, M., Bauer, N. H., Bernloehr, A., ... & Pfeiffer, T. (2020). Project Heb@ AR: Exploring handheld Augmented Reality training to supplement academic midwifery education. DELFI 2020–Die 18. Fachtagung Bildungstechnologien der Gesellschaft für Informatik eV.

[4] Lytridis, C., Tsinakos, A., & Kazanidis, I. (2018). ARTutor—an augmented reality platform for interactive distance learning. Education Sciences, 8(1), 6.

[5] Microsoft (2018) Microsoft HoloLens | The leader in mixed reality technology. https://www.microsoft.com/en-ca/hololens. Accessed 3 Jun 2018

[6] Daqri (2018) Smart Glasses™ - DAQRI. https://daqri.com/products/smart-glasses/. Accessed 3 Jun 2018

[7] Meta (2018) Meta | Augmented Reality. http://www.metavision.com/. Accessed 3 Jun 2018

[8] Leap M Magic Leap. https://www.magicleap.com/. Accessed 6 Jan 2019

[9] Perea, P., Morand, D., & Nigay, L. (2020, September). Target Expansion in Context: the Case of Menu in Handheld Augmented Reality. In Proceedings of the International Conference on Advanced Visual Interfaces (pp. 1-9).

[10] Whitlock, M., Harnner, E., Brubaker, J. R., Kane, S., & Szafir, D. A. (2018, March). Interacting with Distant Objects in Augmented Reality. In 2018 IEEE Conference on Virtual Reality and

3D User Interfaces (VR) (pp. 41-48). IEEE.

[11] Huang, T. K., Yang, C. H., Hsieh, Y. H., Wang, J. C., & Hung, C. C. (2018). Augmented reality(AR) and virtual reality (VR) applied in dentistry. The Kaohsiung journal of medical sciences, 34(4), 243-248.

[12] D. Bowman, E. Kruijff, J. J. LaViola, Jr., and I. P. Poupyrev (2004) 3D User Interfaces: Theory and Practice

[13] Yin, J., Fu, C., Zhang, X., & Liu, T. (2019). Precise Target Selection Techniques in Handheld Augmented Reality Interfaces. IEEE Access, 7, 17663-17674.

[14] Grandi, J. G., Debarba, H. G., Bemdt, I., Nedel, L., & Maciel, A. (2018, March). Design and assessment of a collaborative 3D interaction technique for handheld augmented reality. In 2018 IEEE Conference on Virtual Reality and 3D User Interfaces (VR) (pp. 49-56). IEEE.

[15] El Jamiy, F., & Marsh, R. (2019, May). Distance Estimation In Virtual Reality And Augmented Reality: A Survey. In *2019 IEEE International Conference on Electro Information Technology (EIT)* (pp. 063-068). IEEE.

[16] Goh, E. S., Sunar, M. S., & Ismail, A. W. (2019). 3D object manipulation techniques in handheld mobile augmented reality interface: A review. IEEE Access, 7, 40581-40601.

[17] H. Benko, A. D. Wilson, and P. Baudisch (2006) Precise selection techniques for multi-touch screens, 1263–1272.

[18] Yusof, C. S., Bai, H., Billinghurst, M., & Sunar, M. S. (2016). A review of 3D gesture interaction for handheld augmented reality. *Jurnal Teknologi*, *78*(2-2).

[19] C. Telkenaroglu and T. Capin (2013) Dual-finger 3D interaction techniques for mobile devices, 1551–1572

[20] Unal, M., Bostanci, E., & Sertalp, E. (2020). Distant augmented reality: Bringing a new dimension to user experience using drones. Digital Applications in Archaeology and Cultural Heritage, e00140.

[21] Lilija, K., Pohl, H., Boring, S., & Hornbæk, K. (2019, May). Augmented reality views for occluded interaction. In Proceedings of the 2019 CHI Conference on Human Factors in Computing Systems (pp. 1-12).

[22] Nor'a M N A & Ismail A W (2019) Integrating Virtual Reality and Augmented Reality in a Collaborative User Interface, 59-94

[23] Bai, H., Lee, G., & Billinghurst, M. (2015, December). Free-hand gesture interfaces for an augmented exhibition podium. In *Proceedings of the Annual Meeting of the Australian Special Interest Group for Computer Human Interaction* (pp. 182-186).

[24] Hürst, W., & Van Wezel, C. (2013). Gesture-based interaction via finger tracking for mobile augmented reality. *Multimedia Tools and Applications*, *62*(1), 233-258.

IOP Conf. Series: Materials Science and Engineering **979** (2020) 012010 doi:10.1088/1757-899X/979/1/012010

MR-Deco: Mixed Reality Application for Interior Planning and Designing

F E Fadzli [1] *, A W Ismail [1], R Talib [2], R A Alias [1,3], and Z M Ashari [2]

[1] School of Computing, Faculty of Engineering, Universiti Teknologi Malaysia, 81310 Johor, Malaysia

[2] School of Education, Faculty of Social Sciences and Humanities, Universiti Teknologi Malaysia, 81310 Johor, Malaysia

[3] Azman Hashim International Business School, Universiti Teknologi Malaysia, 81310 Johor, Malaysia

*fazliaty.edora@gmail.com

Abstract. Over the past few years, Augmented Reality (AR) and Virtual Reality (VR) have become a phenomenon in this new era and nowadays smartphone device can enable these technologies quickly and seamlessly through downloadable applications. Some of them are to benefits community and others just for entertainment purposes such as movies and games. Despite their goal, researchers take the opportunity to use Mixed Reality (MR) technology as it is suitable for real-life usage by users as its low cost and it can replace the conventional way of rearranging the furniture for interior design application. MR also brings design and planning, interior design, and engineering into a digital world where people can operate the prototype without having to physically involve in real life. A user should be able to interact intuitively with the virtual content in real-time. This paper identifies the integration for both AR and VR fusion for the MR environment. It also discusses the suitable user interaction technique can be applied in interior design in the MR environment, called MR-Deco.

1. Introduction

Mixed reality (MR) is a technology that either supplement or replace real-life with computer-generated data to facilitate user interaction through the natural senses which including augmented reality (AR) technology and the virtual reality (VR) technology. VR and AR have been useful for exploration, logical training and the development of new models of understanding in complex data systems [1].

As defined by [2], MR in the field of human and computer interactions involves the superposition of virtual reality graphs and enables users to interact with the virtual world tangibly. For the user to have an immersive experience in MR, the virtual scene generated by the computer must be composed with the actual world scene without the sense of incompatibility by three cohorts, which are geometry consistency, consistency of time and optical consistency [3].

Furthermore, for users to have an immersive experience in a VR environment, a user should be able to interact intuitively with the content through three-dimension (3D) space. The application should support several VR environment techniques for interaction and object manipulation such as rotation, translation and user interface (UI). As [4] stated, a key component is the development of appropriate AR application interaction techniques that allow users to interact intuitively with virtual content.

Nowadays, many applications such as games and simulations are using AR and VR, and there is no doubt that these technologies have been on a long journey of research and development. However, still, there is more need to be done to make sure it is reliable, robust and most importantly, convenience and economical to reach a higher market value in this digital era.

As defined by [5], MR is a cutting-edge technology, which connects real and virtual worlds seamlessly. The AR and VR technologies, therefore, naturally immerse users in 3D environments [5], and this may improve the user experience when 6 Degree-of-Freedom (DOF) controller can provide immersive user interaction [6]. ARDeco has demonstrated an interior design for AR environment [7]. Researchers have studied for an advanced user interaction and a comprehensive review has been presented in [8] to show MR has potential for intuitive user interaction. MR interface has proven can enhance user experience depending on the hardware and MR devices [9]. Ray pointing primarily used when the user needs to use HMD with the motion controller, it works similarly with hand rays. Real hand gesture required specific hardware and the accuracy may need to study further in-depth [10]. However, this research chooses to use a ray pointing because the proposed interior design application aims for far pointing in MR. Furthermore, the only reliable MR device nowadays is Hololens, and it is very costly to enable user interaction in such an immersive merged reality environment. Thus, this research attempt to discard the hardware restriction by utilizing the holographic projection as the display technology. Augmented Virtuality (AV) environment works with hologram display has been proposed for MR environment [11]. Therefore, this paper proposes an interior design MR application called MR-Deco that should be able to integrate AR and VR using holographic projection in the MR environment. As agreed by [12], the approaching new advanced technologies have encouraged to implement interior design in MR environment.

2. Interior Design Application Development

Interior designing is chosen as this research application to test the interaction in the MR environment. The interior design application for user interaction and 3D object manipulation such as rotation, translation and also duplicate and remove.

2.1. Defining input controller

This section describes the user interaction in AR and VR technologies for the proposed MR framework. It is essential to identify the integration for both AR and VR fusion for the MR environment.

The proposed prototype for interior design has developed, and user interaction in this prototype used a typical VR interaction. Between real hand gesture and motion controller, as shown in Figure 1, the interaction designs are symmetry. When a user wears an immersive headset and interacts with 3D objects via motion controllers, it will shoot out rays for pointing and manipulating far objects

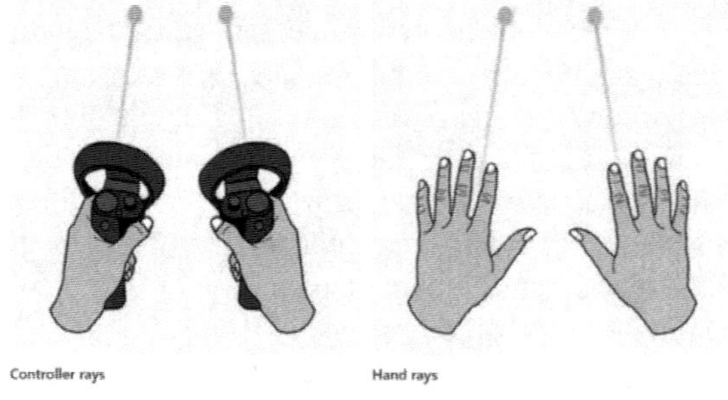

Figure 1. Controller rays and hand rays [13].

. There are buttons on the controllers for further committing different actions. In our proposed MR-Deco, we did leverage the interaction model of rays and attached them to one-handed hand only. Ray pointing is suitable for far pointing and manipulation, primarily when the user uses HMD with the motion controller, it works similarly with hand rays.

The interaction for MR-Deco includes rotation, translation, and also texture manipulation, where the user can change the texture of the object. HMD is essential in this development. There are several interaction controllers can be used with the HMD such as Oculus Touch, Oculus Remote and also Joystick. As illustrated in figure 2 (a), the workspace to set up VR display technology, there are listed the computer desktop, Oculus Sensor, Oculus Rift and Oculus Touch. Oculus Rift is chosen as the display device for users to experience and interact with the MR environment, and Oculus Touch used for the user interaction. Oculus Touch allows users to interact with the virtual object such that their hands are pointing a ray to hit the object (as in figure 2(b)), to be precise the controller rays produce a ray pointing using Oculus's touch controller. Then, for the translation and rotation, the user uses the right-touch controller.

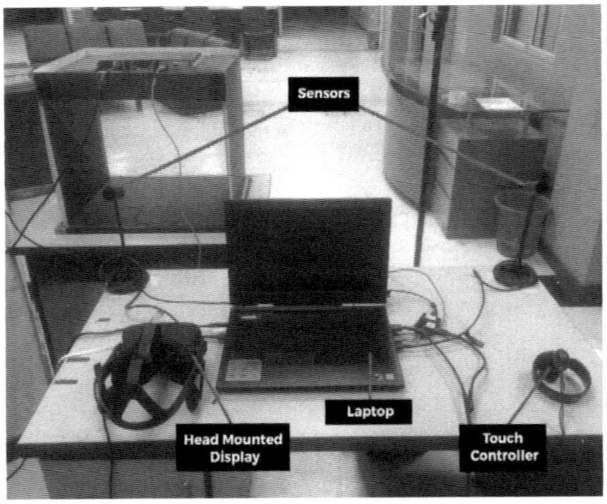

(a) Hardware for workspace setting

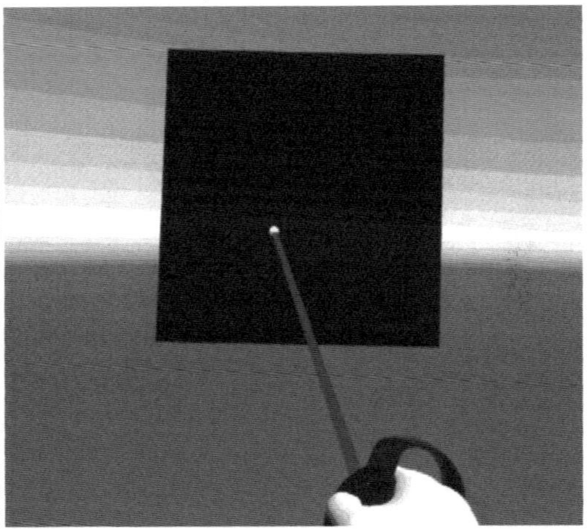

(b) Ray pointing in VR scene using controller ray

Figure 2. VR workspace.

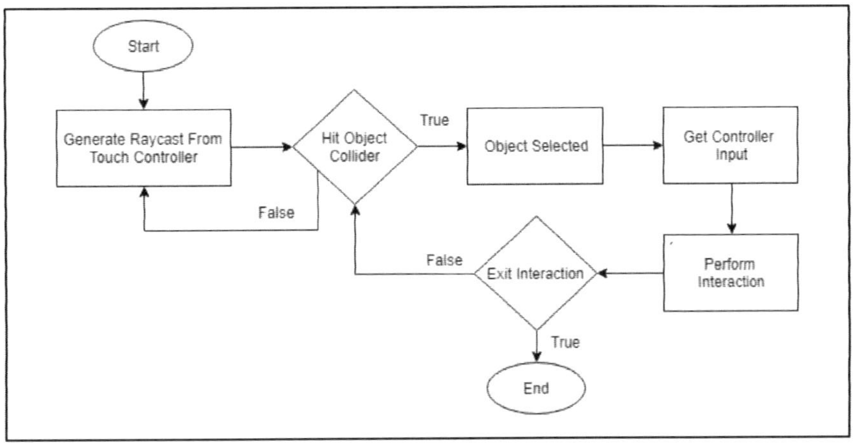

Figure 3. Interaction workflow.

Collision detection has implemented when a ray hits the collider that bounds the object if the collider does not hit the object, the system will return to search for the ray-cast coordinates based on the controller movement as illustrated in figure 3. Once the object is hit, the interaction can be performed. Interaction is typical with the select and the release function. The release will end the interaction; select function will trigger the selected object. There is some crucial component such as ray-cast and collider, to enable interaction between the virtual object and input controller. A button has long been a metaphor used for triggering an event in the 2D abstract world. In the 3D MR world, anything can be an interactable object that triggers an event. Next section will describe the list of interactable objects that have been designed to form a user interface for our proposed MR-Deco application.

2.2. Designing UI for MR-Deco

This application consists of several user interface (UI) designs which each of the UI executes different operation functions such as to open menu panel, to spawn new virtual object and also to remove all the virtual object from the environment. This UI interaction enables the user to customize the environment according to their desire preference. There are many virtual objects in the environment. However, this prototype only focusses on four objects which are furniture, including single seat sofa, double seat sofa, coffee table and torcher lamp, to allow user interaction with the UI design. Those UI designs are as shown in Table 1.

Table 1. Interactable objects.

User Interface	Descriptions
	Button to spawn a new single sofa object in the VR room space.
	Button to spawn a new double sofa object in the VR room space.
	Button to spawn a new coffee table object in the VR room space.
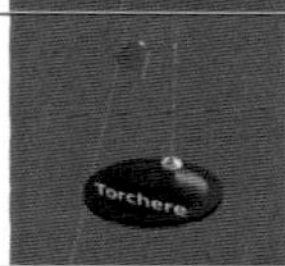	Button to spawn a new torcher object in the VR room space.
	Button to delete all objects in the VR room space.
	Icon for object's properties to duplicate selected object.
	Icon for object's properties to remove selected object.

This application consists of six UI designs which are button to open menu panel, spawn the different type of furniture and also destroy or remove all the virtual object from the virtual environment. To interact with the UI design, the user has to perform a particular gesture to enable the menu panel. There are also object's properties UI for the user to duplicate or remove the selected object.

2.3. Implementing User Interaction
There are three main interactions implemented in this MR-Deco application which includes selection, manipulation and also UI interaction.

2.3.1. Selection
For object selection, the user used a touch controller, and a ray-cast is the crucial component for object manipulation. It allows the user to perform the interaction of object selection on a particular virtual object in the virtual environment and perform the manipulation on the selected object, as shown in figure 4. To complete the selection interaction, user can hover the pointer of the ray-cast to any object which is the furniture and perform the selection by pressing "A" button, as in figure 4 (a). The user has to press "A" button to trigger the selection interaction on the object, and the object change material with a light blue colour as an indicator of the selection has been performed on the selected object, as in figure 4 (b). The ray-cast with a line renderer from the origin of the virtual hand in the virtual environment, as shown in figure 4 (c).

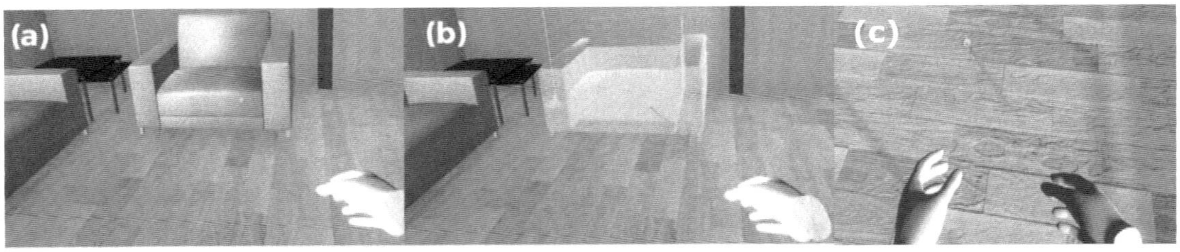

Figure 4. (a) shows the object is deselected in the virtual environment, (b) shows the object is selected in the virtual environment (c) shows a ray-cast with a line renderer for pointing

2.3.2. Manipulation
Based on figure 5, the user can interact with the menu panel and perform the selection. The user can choose any available objects from the listing to spawn a new object into the virtual environment or remove all object from the virtual environment. The user used touch tapping to interact with the menu that consists of an array of objects. There are four initial objects, *Single Seat Sofa*, *Double Seat Sofa*, *Coffee Table,* and *Lamp Torch* have been defined as the interactable objects.

Figure 5. Menu panel to spawn the initial objects

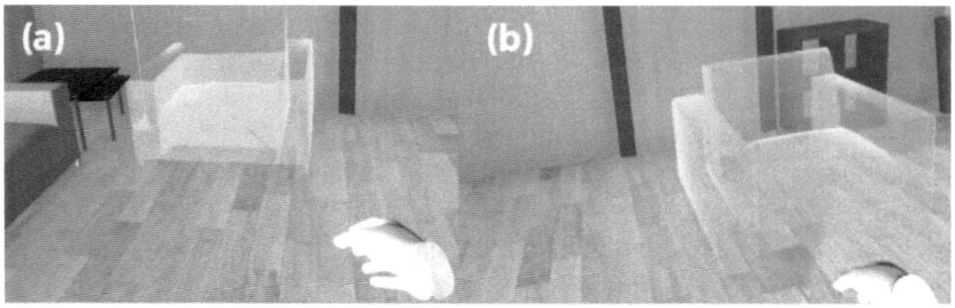

Figure 6. (a) shows the object is selected, (b) shows the selected object has been translated to a new position

Figure 6 shows the user performs the object selection and the selected object has been moved to a new position, translation is performed by hovering or moving the controller to a desired position within the room space. The object can be rotated in clockwise and also anti-clockwise depends on the interaction input of the thumbstick. In order to rotate the selected object clockwise, user can move the thumbstick on Y-axis to the right while moving the thumbstick to the left rotate the selected object anti-clockwise. Thus, the object rotation, as presented in figure 7.

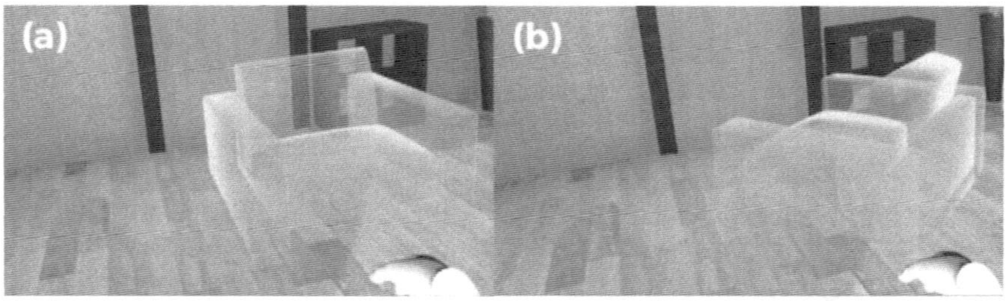

Figure 7. (a) shows the object is selected, (b) shows the selected object has been updated with the new angle for object rotation

2.3.3. Duplicate and Remove

User can perform intuitively using the touch controller the duplicate function on the desired object, and also can remove the selected object. Duplicate event is an alternative way for a user to spawn the same object in the room; the designing activities would be much faster or less time taken. Although there is remove all object function in the main menu panel, the user might also want to delete a specific object that they do not wish to enter the room space. Thus, duplicate and remove functions have designed to affect the selected object. The functions to duplicate and remove the selected object, as shown in figure 8; it is a toggle menu will enable and disable the selected menu.

(a) (b)

Figure 8. shows a toggle menu when the duplicate button is enabled, and the remove button is disabled.

The object's properties menu implemented as UI design interaction technique where the user can duplicate and remove the selected object. The interaction menu can be opened by pressing the "B" button on the Touch Controller and "A" button to close the menu. To select the option in the menu, the user can press the "B" button again to select the option. User can also scroll between those two options in the properties menu by using the thumbstick of the Touch Controller. However, the object's properties menu can only be opened if only an object is selected in the room space of the VR environment. Thus, the results when the user removes and duplicates the object as in figure 9 and figure 10.

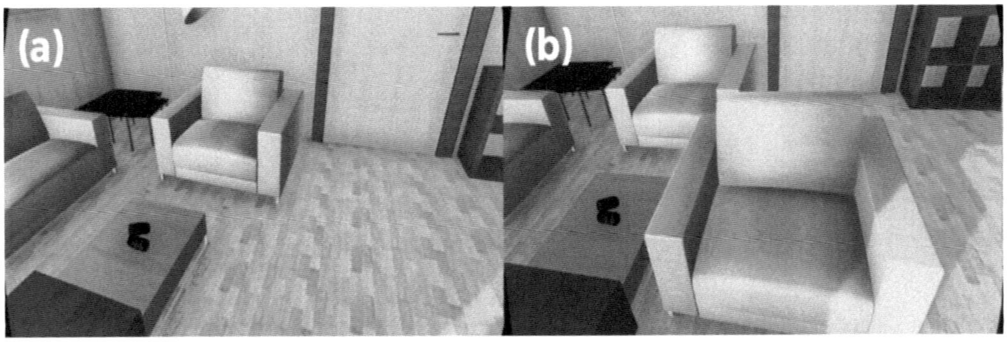

Figure 9. (a) shows the object before the interaction of object duplication while (b) result after the object duplication performed.

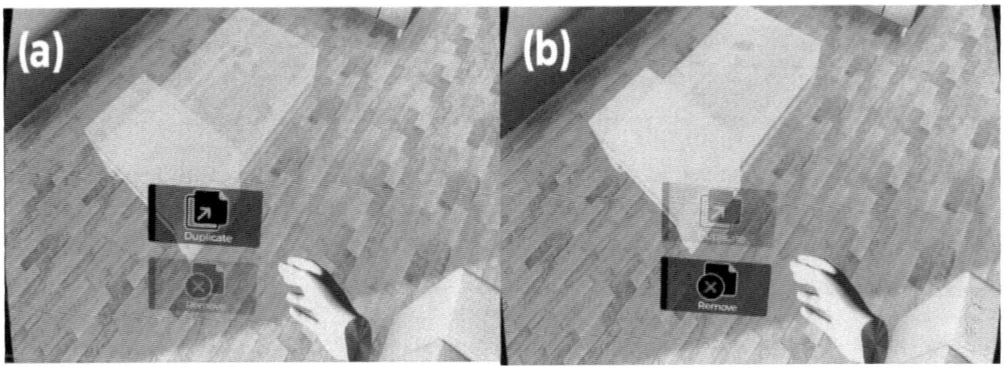

Figure 10. (a) shows the object is selected with the object's properties menu while (b) result after the remove object option is selected.

3. MR-Deco Application in MR

The third phase is to integrate the holographic projection with the Interior Design application in MR. At this phase, it involves the integration the AR holographic projection that corresponds with MR Capture (MRC), Z-hologram and High Definition (HD) external camera is used with the controller to migrate the VR interaction into MR environment.

3.1. Mixed-Reality Capture

VR works with MRC components for this study. MRC needs camera calibration and chroma key composition to be set up manually and accurately to produce a robust camera composition in the MR environment. The workspace has to be set up before the user executes the application such that the camera, Touch Controller and chroma green screen are positions correctly as shown in figure 11.

To enable MRC in the application, a Touch Controller has to set as a VR Object which acts as an anchor of the camera transform properties from the real world into the virtual environment. Then, the camera needs to calibrate using the MRCTools from Oculus SDK. The process of camera calibration captures several frames of the marker provided to produce a virtual camera depth in the VR environment. The calibration pre-defined data has saved as a file and then will be loaded when the MRC plugin has started in the application. Thus, this enables the camera in the virtual environment to move corresponds to the VR Object, which is an external camera attached with Touch Controller.

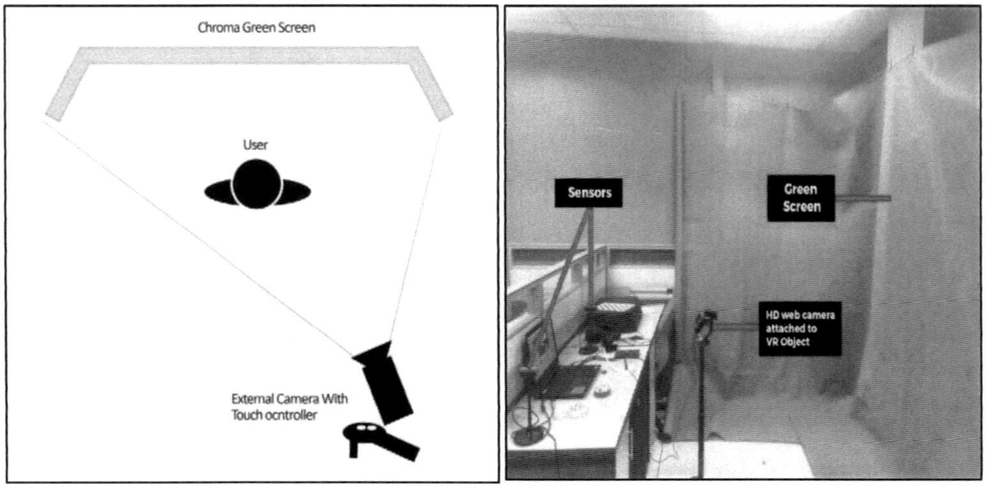

Figure 11. MRC workspace.

3.2. Integrating with the holographic projection

The application is integrated both VR and MRC with AR holographic projection using Z-Hologram for user interaction in the MR environment. In this section, the application should be able to run and execute with the HMD, which is Oculus Rift and use MRC with direct composition while synchronizing interaction with the AR projection in Z-hologram. Users also should be able to interact with the VR object through the interface and perform all the interaction technique included in the application.

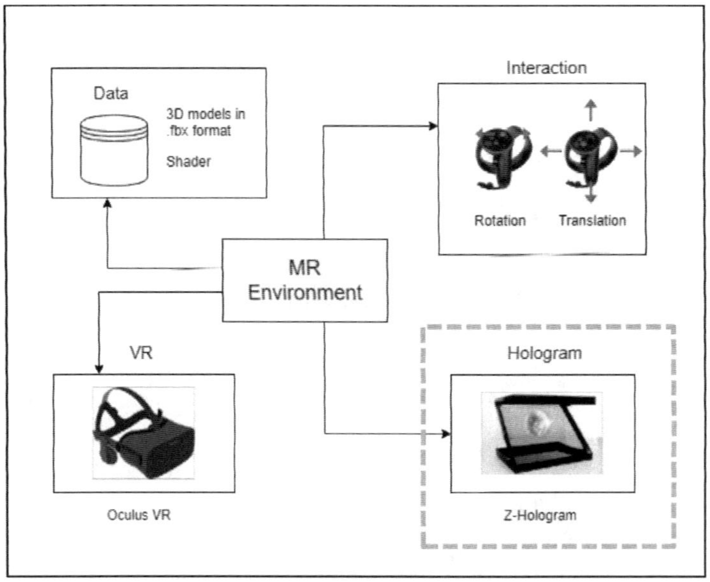

Figure 12. Components in MR environment.

Figure 12 illustrates the MR environment that the user interface consists of several components. There is a data to stores the transmitting data, and the device Oculus VR is to enable VR controller, Z-hologram presents the holographic projection and also the interaction methods such as rotation and translation. The VR object can only be manipulated if and only if the VR object is hit by raycast and selected, as explained in the previous section. MRC also needs to be enabled using direct composition to stream the real-world environment into the virtual environment in real-time. After that, users can interact with the object intuitively through the Oculus Rift controller called Oculus Touch. Therefore, the user feels immersive when interacting with the 3D object in the MR environment.

Based on figure 13, the application system flow describes the process of user interaction in the MR environment. The application enables user interaction in the MR environment by using MRC plugin and updates the interaction in the virtual environment onto the Z-hologram, which augments the object manipulation made in real-time. The integration process involves combing the display technique of VR, MRC and AR holographic projection in a single scene. Thus, MRC display and holographic projection can give the immersive and intuitive experience to the user, and the user should be able to perform the interaction.

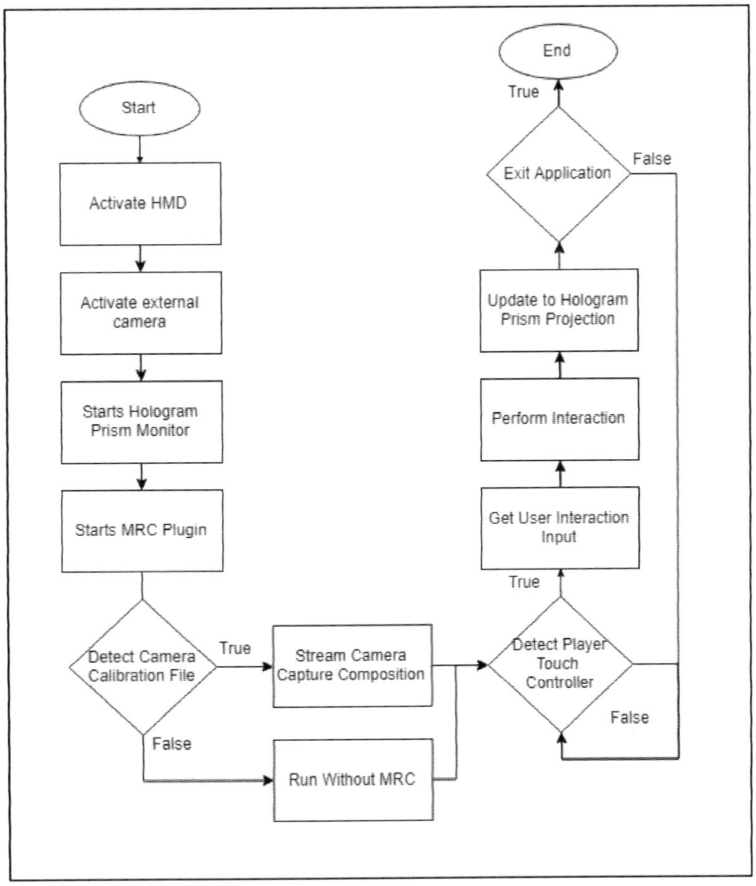

Figure 13. 3D Interior Design application system flow.

4. Results

In the setup, the application requires a laptop to run the application with the Oculus Rift connected to it to authorize the VR component used in the application. Apart from the MRC components, the HD web camera is attached to the Touch Controller and positioned slightly behind the user. The user has to face reasonably opposite to the Oculus set up to give a real impression of the user interacting and being in the virtual environment. Other than that, this application also requires a monitor screen to project the virtual environment using holographic projection which is Z-Hologram to allow spectators to spectate the interaction performed to the virtual objects.

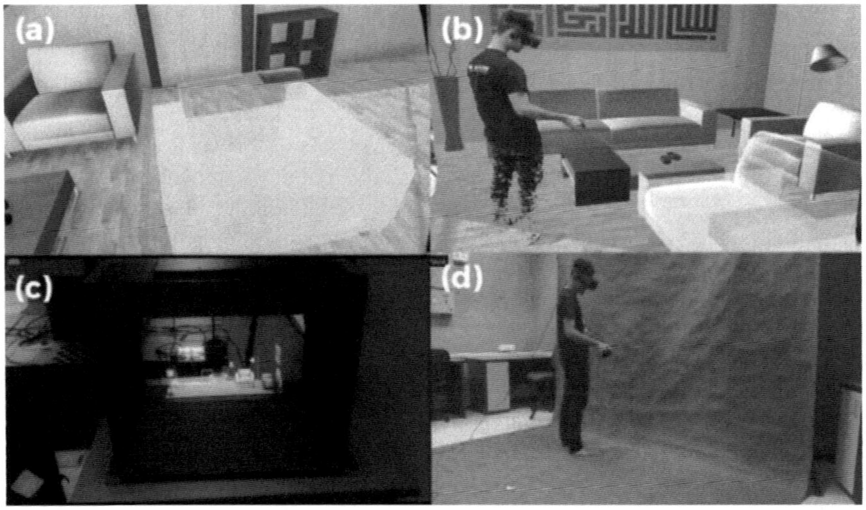

Figure 14. (a) shows the VR display view from the HMD device, (b) shows the MRC display in the laptop display, (c) shows the Z-Hologram display, and (d) actual user's activity in the real world.

The application also integrates the VR interaction and display with the MRC projection display to enable MR environment and also Z-Hologram as holographic projection, as shown in figure 14. Real user exists in VR space, this condition, also known as augmented virtuality (AV). MR encompasses AR and AV, as claimed by [11]. In the real world, as in figure 14 (d)), the user's activity was moving around while holding the controller and user manipulates the VR space (in figure 14 (a)) using the see-through HMD. The participant or non-active users can see the manipulation activities through the holographic Z-projection (as in figure 14 (c)), they can see what the user had modified without needed an HMD.

Figure 15 shows the initial set up for the MR-Deco application; (1) the Oculus is set up facing to the user inside the green screen space set up, (2) the green screen is set up in front of the Oculus setup to enable Mixed-Reality Capture (MRC). (3) the user wears Oculus HMD and holds a right Touch Controller to interact in Virtual Reality (VR) environment. (4) The high definition (HD) web camera is attached to a Left Touch Controller to enable the MRC camera composition and positioned slightly at the side of the Oculus set up to give an intuitive view for the mixed reality (MR) display. (5) The Z-Hologram is set up facing opposite from the Oculus setup and positioned behind the Oculus setup for the user to evaluate the synchronization of the user interaction in VR with the AR holographic projection.

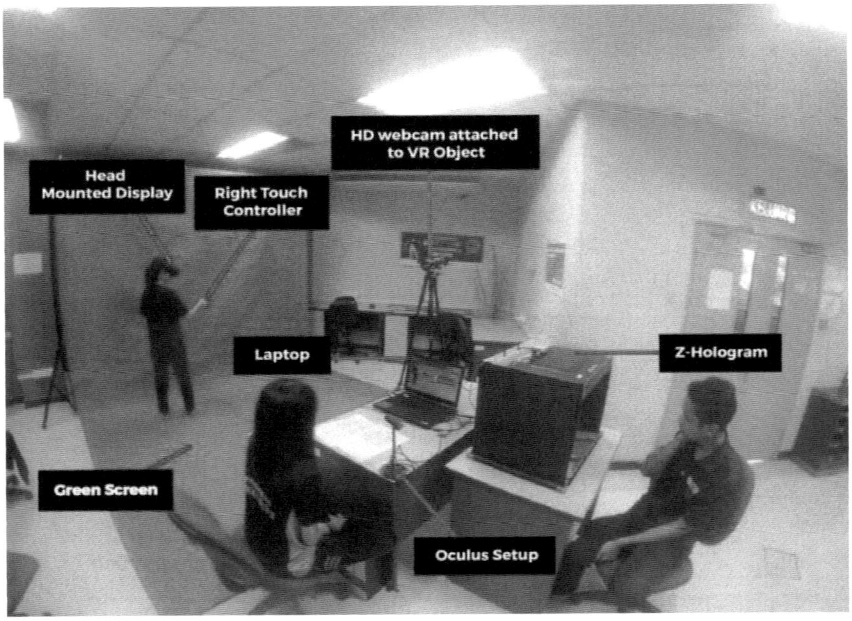

Figure 15. MR-Deco Setup.

5. Conclusion

This paper has discussed the user interaction technique implemented in MR environment for interior designing application called MR-Deco. The interaction input and method applied in the application has been explained as well as the interaction workflow. Throughout this research, there are two limitations faced during the development phase. Firstly, the application can only allow single-hand interaction to utilize only the right Touch Controller rather than both controllers. Thus, the tasks can be performed in a much longer time if compared to both hands used. Other than that, the holographic projection technique used is Z-hologram which not project on all sides and angel in 360 degrees which limit to only the user in front of the hologram who can see the display.

As a suggestion, remote collaboration for multi-user can be the future work to enable multiple users to collaborate in designing the virtual space together in a different set of places remotely. This study can make the application more intuitive and interactive to the users. Besides, user interaction using both hands can be considered instead of just with the right hand so that the tasks can be performed faster and easier by utilizing two Touch Controllers on both hands. Other than that, the holographic projection can use a prism shape as the projection technique rather than Z-Hologram, which enable users to view the projection display from all side and angels.

Acknowledgement
We would like to express our appreciation to Mixed and Virtual Reality Laboratory (mivielab), ViCubeLab at Universiti Teknologi Malaysia (UTM) for the facilities and technical supports.

References

[1] Farahani N, Post R, Duboy J, Ahmed I, Kolowitz B J, Krinchai T, and Pantanowitz L 2016 Exploring virtual reality technology and the Oculus Rift for the examination of digital pathology slides. *Journal of pathology informatics*, **7**.

[2] Qui T C T, Nguyen T H D, Cheok A D, Teo S L, Xu K, Zhou Z, and Le Nam Thang Y L 2010 MagicLand: a 3D Human Capture Mixed Reality System for Museum Experiences. *Re-Thinking Technology in Museums: towards a new understanding of people's experience in*

museums, **71**.
[3] Hachimura K, Kato H, and Tamura H 2004 A prototype dance training support system with motion capture and mixed reality technologies.. *13th IEEE Int. Workshop on Robot and Human Interactive Communication* pp 217-222
[4] Zhou F, Duh H B L and Billinghurst M 2008 Trends in augmented reality tracking, interaction and display: A review of ten years of ISMAR. *In Proc. of the 7th IEEE/ACM International Symposium on Mixed and Augmented Reality* pp 193-202
[5] Piumsomboon T, Lee G A and Billinghurst M 2018 Snow Dome: A Multi-Scale Interaction in Mixed Reality Remote Collaboration *In Extended Abstracts of the 2018 CHI Conf. on Human Factors in Computing Systems* p D115
[6] Ismail A W, Billinghurst M, Sunar M S, and Yusof C S 2018 Designing an augmented reality multimodal interface for 6DOF manipulation techniques. *In Proc. of SAI Intelligent Systems Conf.* pp 309-322
[7] SyafiqahSafiee N and Ismail A W 2018 AR home deco: virtual object manipulation technique using the hand gesture in augmented reality *Innovations in Computing Technology and Applications*
[8] Aladdin M Y F and Ismail A W 2019 A review on multimodal interaction in Mixed Reality Environment *In IOP Conf Series: Materials Science and Engineering* (Vol 551, No 1, p 012049)
[9] Aladin M Y F and Ismail A W 2019 Designing User Interaction using Gesture and Speech for Mixed Reality Interface *Int. Journal of Innovative Computing*, 9(2).
[10] Fadzli F E.and Ismail A W 2019 VoxAR: 3D Modelling Editor Using Real Hands Gesture for Augmented Reality *In 2019 IEEE 7th Conf. on Systems, Process and Control (ICSPC)* pp 242-247
[11] Milgram P and Colquhoun H. 1999 A taxonomy of real and virtual world display integration. Mixed reality: Merging real and virtual worlds, **1**, pp 1-26.
[12] Janusz J 2019 Toward the new mixed reality environment for interior design. *In IOP Conf. Ser. Mater. Sci. Eng* (Vol 471, p 102065)
[13] Retrieved from https://docs.microsoft.com/en-us/windows/mixed-reality/.
[14] Nor' a M N A and Ismail A W 2019 Integrating Virtual Reality and Augmented Reality in a Collaborative User Interface *Int. Journal of Innovative Computing* 9(2).
[15] Ismail A W 2019 A review of the collaborative learning environment across virtual and augmented reality technology *In IOP Conf. Series: Materials Science and Engineering* (Vol 551, No 1, p 012050)
[16] Fadzli F E, Ismail A W, Aladin M Y F and Othman N Z S 2020 A Review of Mixed Reality Telepresence. *In IOP Conf. Series: Materials Science and Engineering* (Vol 864, No 1, p 012081)

IOP Conf. Series: Materials Science and Engineering **979** (2020) 012011 doi:10.1088/1757-899X/979/1/012011

AR-TO-KID: A speech-enabled augmented reality to engage preschool children in pronunciation learning

M Y F Aladin [1,2] *, A W Ismail [1,2], M S H Salam [1,2], R Kumoi [1,2] and A F Ali [1,2]

[1] ViCubeLab Research Group, Universiti Teknologi Malaysia, 81310 Johore, Malaysia
[2] School of Computing, Faculty of Engineering, Universiti Teknologi Malaysia, 81310 UTM Johor Bahru, Johor, Malaysia

* yahyfekri@gmail.com

Abstract. AR-TO-KID is an application produced for preschool children between ages five to six years old with an Augmented Reality (AR) application. The significant purpose of AR-TO-KID is to improve the pronunciation of the children in English. Hence, this paper discusses an AR application with speech input. The detection of the children speech input when they need to pronounce the words correctly, and they need to have critical thinking to identify the environment suit with the 3D objects that they will utter the word. Educational technology should be interactive and attractive for 5 to 6 years old preschool children learning; however, some at preschool teachers still used the conventional methods in teaching and children are not fully engaged with the method. Therefore, this project is to design and develop an interactive AR tool called AR-TO-KID for preschool children in pronunciation learning and teaching. This paper presents the evaluation and testing for preschool children with non-native English speaking. The article ends with results and discussion.

1. Introduction

In Malaysia, speaking in English as a second language is an excellent way to get a grasp of the language. We believe that speech is good at traversing interaction barrier, especially to young children because it lets them cut through reading difficulties and diffidence with a natural or more familiar way of communicating in their age [1]. However, minimal use of the language at home or in school has made non-native young children feel less motivated to use the language [2]. Speech recognition can be useful for the purpose of language learning, where it can teach proper pronunciation and help young language learners develop fluency with their speaking skills. This paper explores how Augmented Reality (AR) and speech recognition technologies could be used together for English pronunciation learning with children that are not native English speakers. We aim to examine the potential of AR to work with speech-enabled system for English pronunciation learning to pre-school children in Malaysia who do not speak English as their first language.

AR is a technology that required a tracking system to bring the virtual content appears on the top of the real-world [3]. Mobile or smartphones are capable to execute AR application, the camera can be a display device to capture the targeted features or known as a marker. The marker is a unique pattern that used to store or register the AR content. The AR tracking process is taking on a marker detection and a marker recognition [4]. The marker must to remain perceived by the AR application otherwise the AR content will be disappeared [4]. User interaction using speech input is usually used with gestures in AR [5][6]. Speech recognition requires the list of grammar to be retrieved by the AR systems.

The proposed idea is the mobile AR application with speech-enabled system to be more significant to children 5-6 years old as it involves logical thinking for the kid to acknowledge the right and wrong based on the situation [7]. The concept of the real problem applies in AR to create a fun learning environment. Therefore, this paper will assess the effectiveness of using AR for English pronunciation learning, and it also discovers if combining speech input with AR cues facilitates the non-native English speaking children to learn English pronunciation. The main contributions of the paper include insights into how speech input with AR can enhance the children's interaction with an AR application and influence their learning experience. The more that is known about the effectiveness of speech input and AR for non-native young children's English language learning, the more effective learning approaches can be developed [8][9].

2. Methodology

The primary purpose of this application is the use of speech recognition with AR implementation. The proposed method has been described in the methodology as presented in Figure 1.

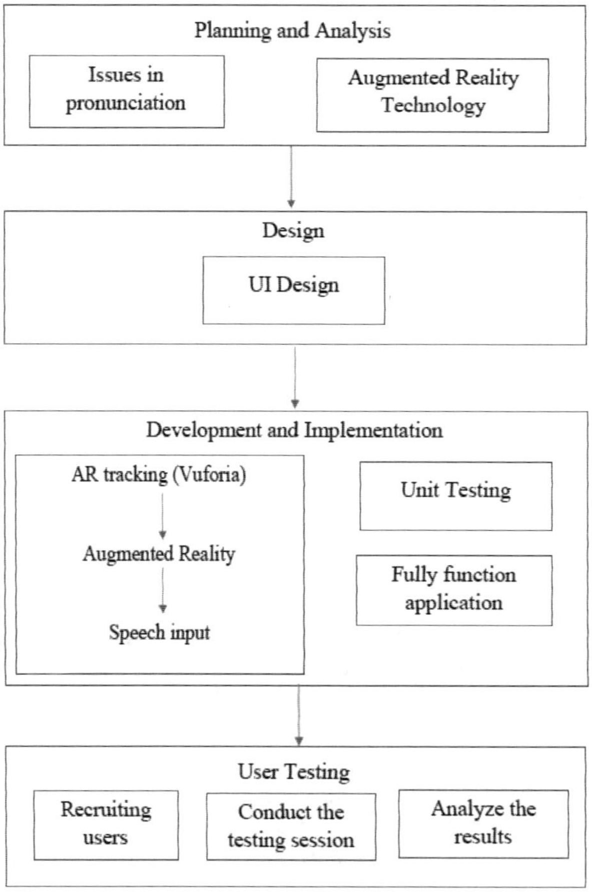

Figure 1. Research Methodology.

Figure 1 shows the methodology as a guideline to this development stages. There are four phases in this methodology which are Planning and Analysis phase, Design phase, Development and Implementation phase and User Testing phase. These phases are the phases that we followed in developing our application—visiting the preschool community to collect data and to know the issues that may occur during the learning lesson. The issue that arises is the kids have problems with their pronunciation. After discussion, the solution for this issue is by making an AR application where the kids will pronounce the correct word that matches the object display to make the object (answer) appear. Through this project, we are using the Unity 2018 software to create the 3D object to apply the AR. Therefore, we designed the UI design using the same software. In the third phase, development is to enable AR by using the Vuforia, tracking for the application related to the AR and with the speech input. The fourth phase is about user testing of the application. In this phase, the application will be tested by the user for the evaluation of the application. Our users were preschool children with age between 5 to 6 years old.

2.1. Phase 1: Designing the User Interface (UI).

The usage of speech recognition is to focus on improving the pronunciation of the children as well as their logical thinking. The application requires the kid to speak the answer based on their logic when they understand the AR objects scene that appeared after the scanning on the marker. That is the reason why this application includes to improve their logical thinking. The right answer will pop out in a 3D shape after the user speaks the correct answer with the correct pronunciation. There will be three different 3D objects will show up on each scenes when the user plays the application.

In workspace setup as in Figure 2, the children were sitting on the chair while holding the smartphone and hover the camera to the marker. The marker is put on the table. When the children hover the camera to the marker, the 3D object will appear on the screen. Speech was recorded in classroom, the classroom environment sometimes produce unneccessary noise, and it has overlapped the speech of the children when they are speaking in the microphone. However, by using handheld device with the built-in microphone, the utterance more clear if the user sits close enough with the device.

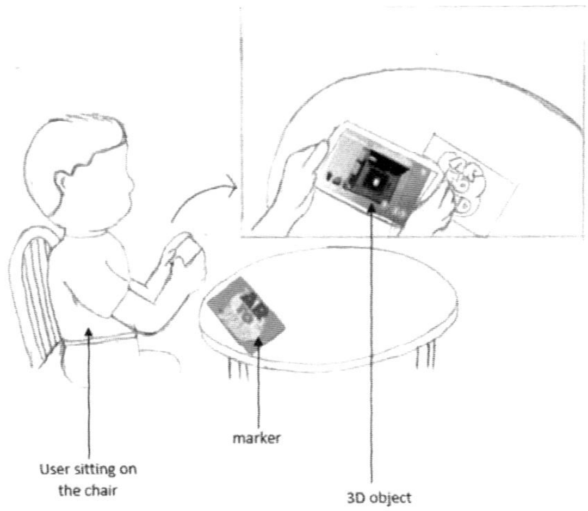

Figure 2. AR-TO-KIT Workspace.

Once the application starts, the camera with scan the marker and pop up question panel appear on the screen. The questions are different following each of the scene created for this application. The user can click on the 'X' button on the top right of the question panel once they have understood the question to close the panel (as in Figure 3 (a)). The grape, rock and vase images displayed on the screen are the options for the user to choose the answer. The user can hear the correct pronunciation of each of the solutions by clicking those images. The 'Speech Detection' on the top of the screen is for the user to know the words that the application detect. If the user taps on the 'Setting' button that located on the top right of the screen, it will display the setting panel, as shown in Figure 3 (b). There are three buttons on the panel, which are the 'Question' button, 'Instruction' button and 'Home' button. When the user

IOP Conf. Series: Materials Science and Engineering **979** (2020) 012011 doi:10.1088/1757-899X/979/1/012011

clicks on the 'Question' button, it will display back the question panel, as shown in Figure 3 (a). If the user hits the 'Instruction' button, it will display a pop-up instruction page, as shown in Figure 3 (c). When the user tap on the 'Home' button, it will return to display the home screen, as shown in Figure 3 (a). The user can tap on the 'X' button to close the setting panel (as in Figure 3 (b)).

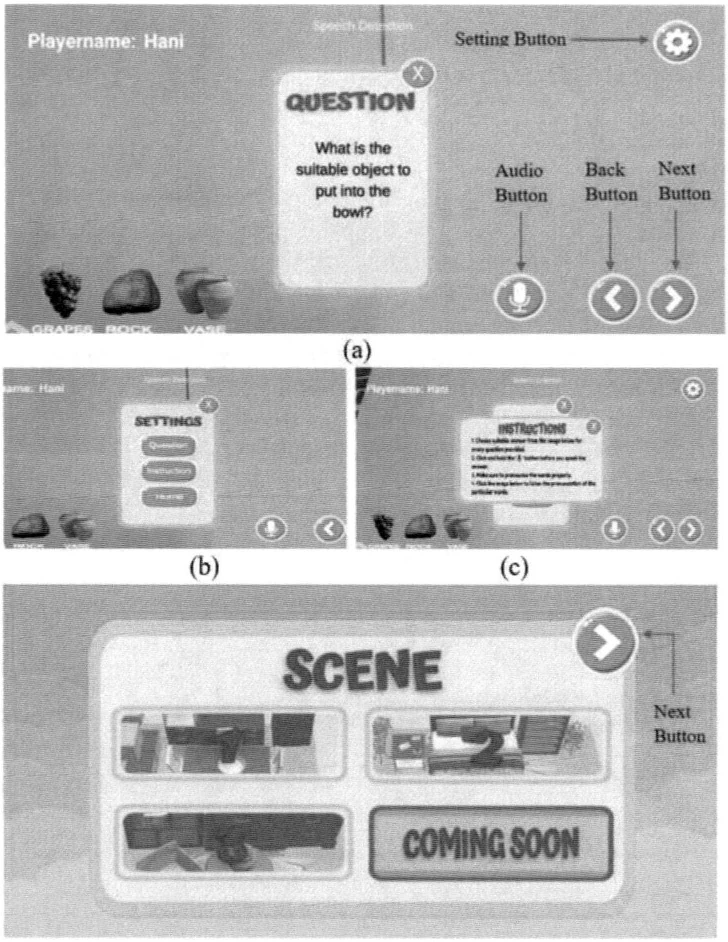

(d)

Figure 3. User Interface Design.

Figure 3 (d) shows the scenes available in this application. The 'Coming Soon' button is to show the user that there will be more scenes coming soon. The number on each of the button shows the number of the scene provided in the application. When the user clicks on the first scene that indicates '1', it will display the first scene, as shown in Figure 4 (a). If the user clicks on the second scene that indicates '2', it will display the second scene, as discovered in Figure 4 (b) and for the third scene that indicates '3', the last scene will be displayed as shown in Figure 4 (c). If the user clicks on the 'Next' button, it will go back to the home screen, as shown in Figure 3 (d).

(a) Scene 1 (b) Scene 2 (c) Scene 3

Figure 4. 3D Scene appears on the marker once the scene is selected.

2.2. Phase 2: Enabling AR Tracking

AR is an interactive experience of a real-world environment where the objects that live in the real world are enhanced by computer-generated perceptual information, sometimes across multiple sensory modalities. In this project, we develop the interactive AR application named as AR-TO-KID by using Unity 3D version 2018 as a rendering engine with the Vuforia SDK for the tracking system to convert our marker into a feature.

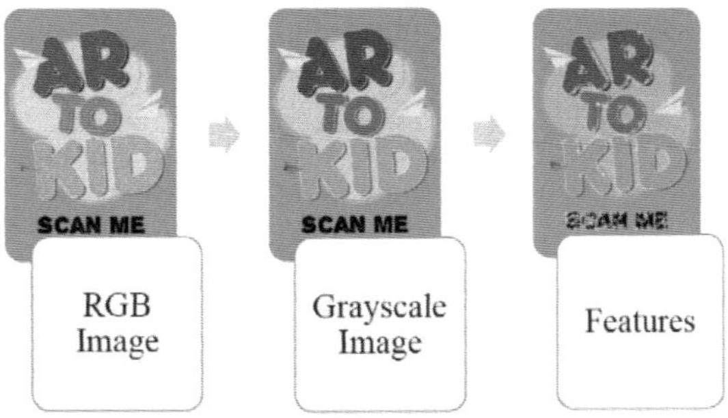

Figure 5. Creating a marker using Vuforia SDK.

Figure 5 shows our RGB marker has been converted into features, we used a single marker, coloured image in RGB as our image target and uploaded it into the database in the Vuforia SDK. This features-based tracking system will convert our image target from RGB to grayscale version to identify the features that can be used for recognition and tracking.

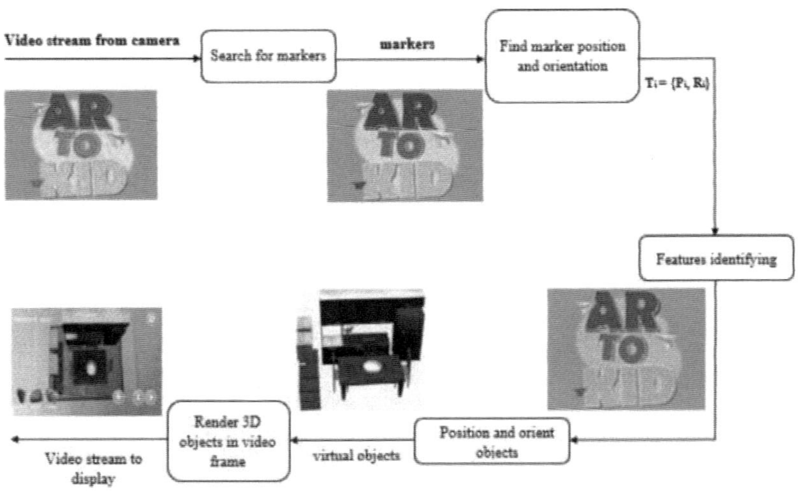

Figure 6. Features-Based Tracking Process for AR-TO-KID.

IOP Conf. Series: Materials Science and Engineering **979** (2020) 012011 doi:10.1088/1757-899X/979/1/012011

The recognition process is presented in the figure 6, the marker is defined as the image target. We improved the performance of our target by enhancing the visibility of these features through adjustments to our target's design, the rendering and scale of our target and how it is printed. Image target represents the image that the Vuforia Engine can detect and track. The image is converted to the binary image, as shown in Figure 6 and find the position and orientation of the image. The image positions and orientations relative to the camera are calculated using Ti = {Pi, Ri}. Then, the Engine detects and tracks the features that are naturally found in the image by comparing these natural features with a known target resource database. By using Ti, the 3D virtual objects are transformed to align them with the marker and rendered them in the video frame.

2.3. *Phase 3: Speech-Enabled System with AR*

For speech input, we were using IBM Watson Speech to Text service on the IBM Cloud, which this service will convert the human voice into written text. It provides speech transcription capabilities for our application. It continuously updates and restores its dictation as it accepts more voice inputs.

In AR-TO-KID application, we have decided to use this service to make our AR object moves using speech. The list of grammars used in this system was:

a. Scene_1_Grammar: "grapes", "rock", "vase"
b. Scene_2_Grammar: "ball", "teddy bear", "television"
c. Scene_3_Grammar: "flower", "lollipop", "pencil"

Table 1. List of the English Pronunciation.

Scene and Questions	Pronunciation words
Scene 1: as in Figure 3 (a)	Grapes, Rock and Vase
Question: *What is the suitable object to place on the bowl?*	Answer: Grapes
Scene 2:as in Figure 3 (b)	Ball, Teddy Bear and Television
Question: *What is the suitable object to put on your bed?*	Answer: Teddy Bear
Scene 3: as in Figure 3 (c)	Flower, Lollipop and Pencil
Question: *What is the suitable object to put into the vase?*	Answer: Flower

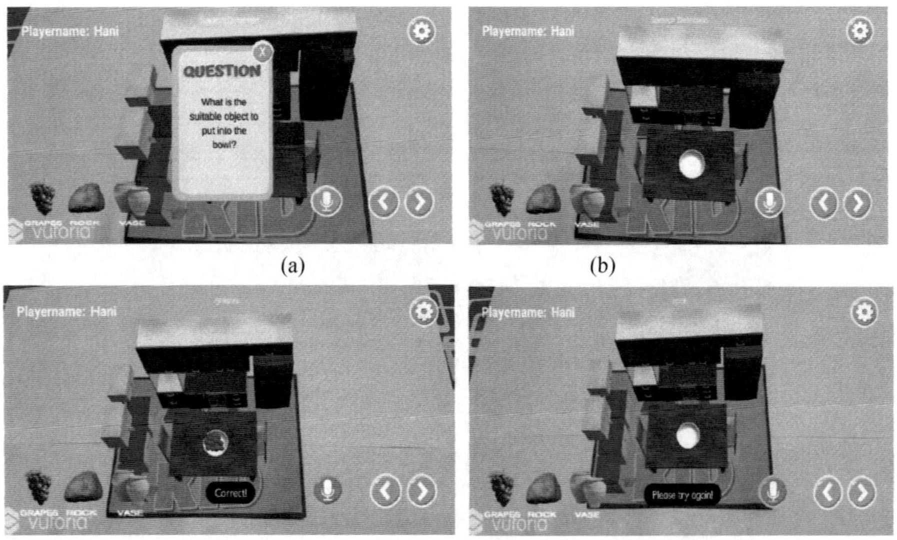

(a) (b) (c) (d)

Figure 7. AR-TO-KID works with speech.

The scene, as shown in Figure 7 (a) will appear the question panel, and the AR-TO-KID will play the wave to sound "What is the suitable object to place on the bowl?". After the children close the panel, there are three images act as a 2D button that positioned linearly at the bottom left of the screen are the Grapes, Rock and Vase (as in Figure 7 (b)). These buttons represent the answer options for the user to choose from. The children can click on those buttons to hear the pronunciation of those particular icons. Then the children will need to say based on the selected object with the correct pronunciation, and the children have to touch and holds the 'Audio' button. The 3D object which is the grapes will fill the bowl, and a 'Correct' word will appear on the bottom of the screen once the children say the correct answer as shown in Figure 7 (c). If the children say the wrong answer or the other words besides the right answer, the screen will display 'Please try again!' statement, as shown in Figure 7 (d).

3. Results and Evaluation

As agreed by [12], there are no studies in the literature which define most appropriate usability guidelines for designing mobile-based AR applications for kindergarten kids. Age is a significant factor to be considered in developing applications for kids, especially when it comes to interaction techniques to be used in learning applications. The kids may not readily use and interact with the applications developed for adults. As a pre-test, the children were advised to fill out a short questionnaire before exploring the application. From the pre-test questionnaire, their background information including knowledge competency of English pronunciation were collected and we also get the teacher feedbacks. After that the post-test had carried out, they were asked the questions as listed in Table 2.

Table 2. List of Tasks in AR-TO-KID.

Tasks
1. User read the instructions by clicking the 'Instructions' button before the start.
2. User clicks the 'Credit' button.
3. User clicks the 'Start' button.
4. The user inserts their name before click 'Enter' button.
5. User clicks 'Enter' button.
6. The user hovers the camera to the marker.
7. User clicks on the 'X' button on a pop-up question after reading it.
8. User clicks on the image displayed on the screen to hear the pronunciation.
9. The user speaks the answer while click and hold the 'Audio' button.
10. User clicks on the 'Setting' button.
11. User clicks on the 'Question' button on setting panel to read the question again.
12. User clicks on the 'Instructions' button on setting panel to read the instructions again.
13. User clicks on the 'Home' button on setting panel.
14. User clicks on 'Next' button.
15. User clicks on the 'Back' button.
16. User clicks on the available scene '1' to '3' button on the 'Scene' page.
17. User clicks on the 'Next' button on 'Scene' page.
18. User clicks on 'Quit' button.

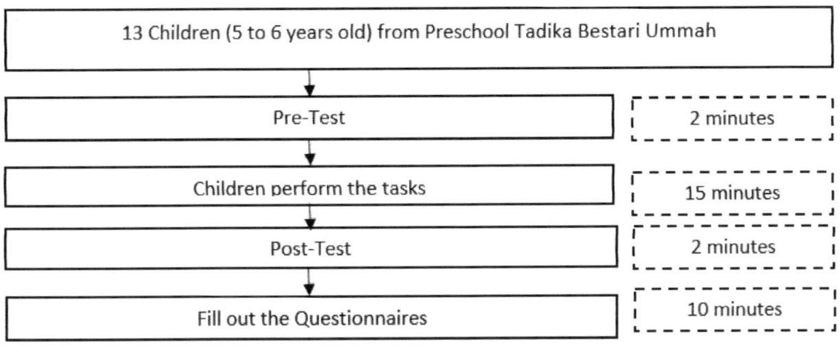

Figure 8. The experiment workflow.

IOP Conf. Series: Materials Science and Engineering **979** (2020) 012011 doi:10.1088/1757-899X/979/1/012011

At the end of the experiments, the children answered the questionnaire by using the smiley Likert scale (as in figure 8). Figure 8 shows the experiment workflow where the 13 children will run pre-test and post-test.

The learning session was conducted at the two kindergarten schools, their school will ensure that the children felt comfortable because they familiar with the environment, they will give more attention and cooperated well without feeling hesitant, this suggestion according to [11]. Thirteen children become our respondents. The learning session during the experiments has been recorded, the children hold the handheld device and hovers to the marker that found on the table, as shown in Figure 9.

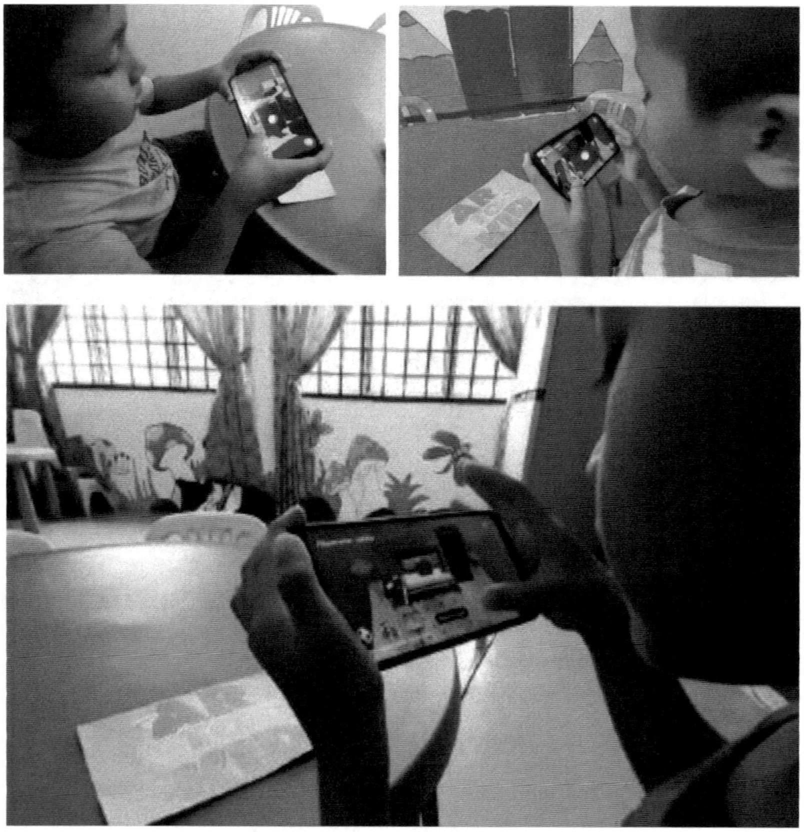

Figure 9. Children play with AR-TO-KID application.

Figure 10. 7-Likert scale in AR-TO-KID for kids' usability ratings.

We used the smiley Likert scale for usability ratings, based on [13]. These emoji design used to replace the rank 1-7 Likert scale for children to understand the questions and straightforward answer the question. The scale in smiley style was redesigned to appear more colourful so it can suit the children to indicate their feeling. This list of questionnaires has been modified based on [13].

Table 3. List of the questionnaires [13].

#	Questionnaires
Q1	I enjoyed doing this activity very much
Q2	This activity was fun to do
Q3	I did not feel boring with this activity
Q4	This activity hold my attention at all
Q5	I would describe this activity as very interesting
Q6	I thought this activity was to improve my pronunciation
Q7	While I was doing this activity, I was thinking about how much I enjoyed it

Four children choose 5 and 7 scale for Q1. It is because they can pronounce the word better than others. The graph in Figure 11 shows that there are the children who have rated 3 and 4 where they did not enjoy the activity because the application has been interrupted by unstable internet. They think input voice is not really the best idea due to it required strong internet connection and also the application needs the quiet workspace. In Figure 12, two of the children choose to rate 5, 8 children have chosen 6, and 3 children have chosen 7. The children have fun with the application when they can see the 3D object appears on the top of the marker, and the application understand their voice command.

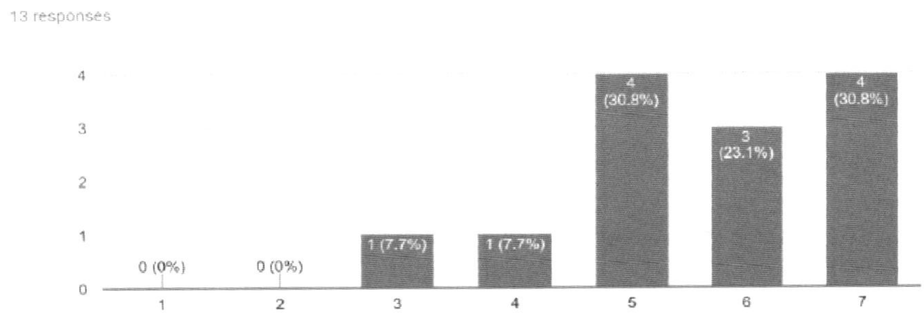

Figure 11. Response to the children enjoy this activity very much.

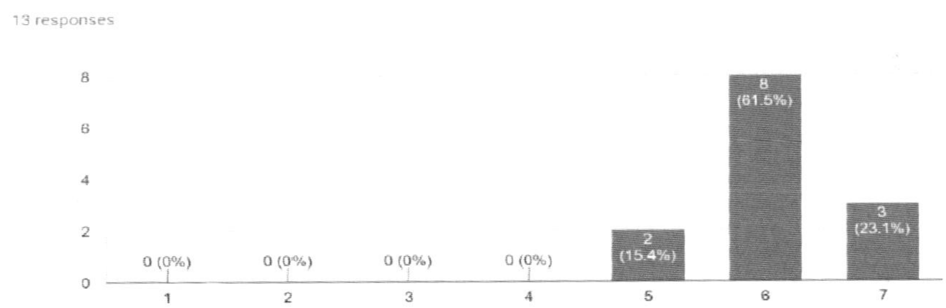

Figure 12. Response to the children found this activity is fun.

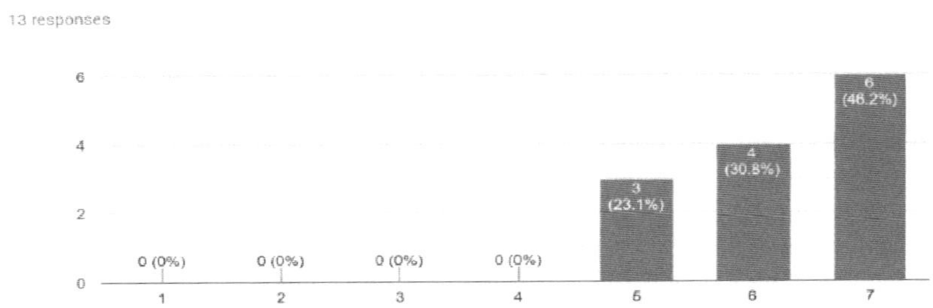

Figure 13. Response to the children found this activity is not boring.

Figure 13 has proven six children strongly agree where the application is not boring, while seven of them feel moderate. It is because the children understand the task they need to complete and they found AR is attractive and when the application brings out the voices for the question they have selected, it likes a virtual teacher has spoken to them.

Figure 14 shows how AR-TO-KID can get their attention. Only one participant has rated 5. Seven children have agreed the application hold their attention. Five children strongly agreed they can pay attention to the AR application.

Figure 15 children strongly agree where the application is impressive. The children think the application is interesting because they have to interact with the application using their voice. There is only one respondent who chooses four which is moderate. It is because the main task of the application, which is using input voice is quite hard due to demanding a strong internet connection and a quiet place. The application might not detect his/her voice during the testing session caused the user not really interested in the application.

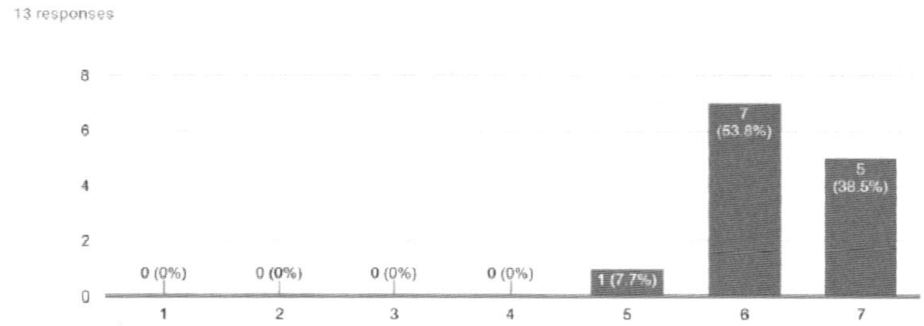

Figure 14. Response to the children found this activity hold their attention.

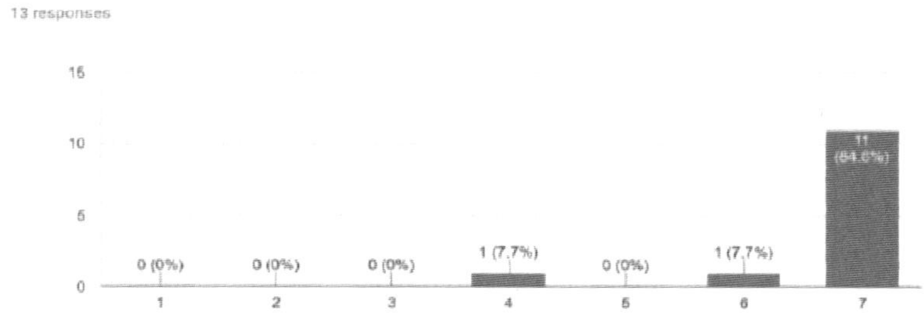

Figure 15. Response to the children found this activity very interesting.

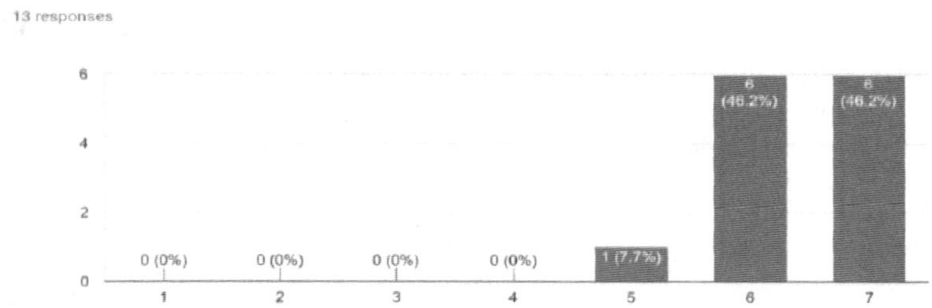

Figure 16. Response to the children found this activity improve their pronunciation in English.

Figure 16 shows the children strongly agree where the application helps to improve their pronunciation in English even they found it difficult to say it right because they are non-native English children. The children think the application has corrected their English when it comes to Scene 3. They have to interact with the application using their voice; they have to say the correct word to complete the task. Only one participant feel disappointed with the application; the children have tried many times but still failed.

Based on the observation results, the participants who are the preschool children are non-native English. The application run the pronunciation must in the British accent. If the user says the word in an American accent or Malay accent, the application might have problems in detecting the user's pronunciation. The other issues arise; the application is an internet-based that require an internet connection to activate the speech recognition or otherwise, it will be not working. Therefore, a stable network connection is required during the testing. The feedback also, if the user says the word without holding the 'Speak' button, the application will not recognize the input, and it does not work. The environment with noisy too much voices will disturb the application. We need to ensure the application must be used in a quiet environment if they want to acquire high effectiveness of the application. If it is in the noisy surrounding, the application will have difficulty to detect the user's voice as there will be too many voice input in the background.

4. Conclusion

The primary purpose of this application is the use of speech recognition with AR implementation. The usage of speech recognition is to focus on improving the pronunciation of the children ages 5-6 years old as well as their logical thinking. The application requires the kid to speak the answer based on their logic when they understand the AR objects scene that appeared after the scanning on the marker. Marker detection is interrupting due to the light, condition of the classroom. The learning session was conducted at the two kindergarten schools, their school will ensure that the children felt comfortable because they familiar with the environment, they will give more attention and cooperated well without feeling hesitant. AR application using speech has improved preschool children pronunciation issues. The children can feel enjoy, and the AR application has improved the engagement. Children are easy to feel boring, AR is attractive and exciting for them, which could hold their attention.

Future works have been identified, and the speech recognition should be built-in with the application to avoid interruption due to the network coverage. The marker should be replaced with the more features extraction; it was suggested to replace the marker that contains more features to produce a robust marker recognition and overcome lighting disturbance [14]. Another suggestion, the AR-TO-KID application can use the marker-based interaction so children more feel realistic with the tangible interaction where they can hold the marker to interact [14]. The learning application also can be supported or assisted by invoke a conversational storytelling agent [15].

Acknowledgement

We would like to express our appreciation to Mixed and Virtual Reality Laboratory (mivielab), ViCubeLab at Universiti Teknologi Malaysia (UTM) for the facilities and technical supports. We also would like to thank to Pre-school Kindergarten, *Tadika Bestari Ummah* and *Tadika Al Ummah* which located in Johor Bahru for the user testing.

References

[1] Azman, H. (2016). Implementation and Challenges of English Language Education Reform in Malaysian Primary Schools. 3L: Language, Linguistics, Literature, 22, 65-78. https://doi.org/10.17576/3L-2016-2203-05

[2] Ansawi, B. (2017). Promoting the 3Es (Exposure, Experience, Engagement) in an English-Rich Rural Primary School Community. The English Teacher, 46, 30-42.

[3] Ismail, A., & Noh, Z. (2008). Augmented Reality Theory and Applications. M. Shahrizal, & M. Najib, Advances in Computer Graphics and Virtual Environment, 88-105.

[4] Ismail, A. W., Billinghurst, M., & Sunar, M. S. (2015, August). Vision-Based Technique and Issues for Multimodal Interaction in Augmented Reality. In Proceedings of the 8th International Symposium on Visual Information Communication and Interaction (pp. 75-82). ACM

IOP Conf. Series: Materials Science and Engineering **979** (2020) 012011 doi:10.1088/1757-899X/979/1/012011

[5] Ismail, A. W., & Sunar, M. S. (2015). Multimodal fusion: gesture and speech input in augmented reality environment. In Computational Intelligence in Information Systems (pp. 245-254). Springer, Cham.

[6] Aladin, M. Y. F., & Ismail, A. W. (2019, August). A review on multimodal interaction in Mixed Reality Environment. In IOP Conference Series: Materials Science and Engineering (Vol. 551, No. 1, p. 012049). IOP Publishing.

[7] Palupi, A. N. (2020). Use of Manipulative Media as A Stimulation of Ability to Understand The Concept of Early Children's Age. Early Childhood Research Journal (ECRJ), 2(1), 43-66.

[8] Potamianos, A. and Narayanan, S. (2003). Robust Recognition of Children's Speech. IEEE Transactions on SAP, 11(6):603–615, Nov

[9] Matassoni, Gretter, R., Falavigna, D., and Giuliani, D. (2018). Non-native children speech recognition through transfer learning. In Proc. of ICASSP.

[10] Eskenazi, M. (2009). An overview of spoken language technology for education. speech communication. Speech Communication, 51(10):2862–2873.

[11] Mispa, K., Mansor, E. I., Kamaruddin, A., & Hinds, J. (2016). Measuring usability and children's enjoyment of virtual toy in an imaginative play setting: A preliminary study. Journal of Theoretical and Applied Information Technology, 89(1), 45-52.

[12] Tuli, N., & Mantri, A. (2020). Usability Principles for Augmented Reality based Kindergarten Applications. Procedia Computer Science, 172, 679-687.

[13] Hall, L., Hume, C., & Tazzyman, S. (2016, June). Five degrees of happiness: Effective smiley face likert scales for evaluating with children. In Proceedings of the The 15th International Conference on Interaction Design and Children (pp. 311-321).

[14] Ismail, A. W. (2011). User Interaction Technique With 3D Object Manipulation in Augmented Reality Environment (Doctoral dissertation, Universiti Teknologi Malaysia).

[15] Ureta, J., Brito, C. I., Dy, J. B., Santos, K. A., Villaluna, W., & Ong, E. (2020, September). At Home with Alexa: A Tale of Two Conversational Agents. In International Conference on Text, Speech, and Dialogue (pp. 495-503). Springer, Cham.

Educational science game for early primary school

A H A Latif [1], M S M Rahim [1,2], A W Ismail [1]*, N M Suaib [1], M Y F Aladin [1], and M N A Nor'a [1]

[1] School of Computing, Faculty of Engineering, Universiti Teknologi Malaysia, 81310 UTM Skudai, Johor, Malaysia.
[2] Media and Game Innovation Centre of Excellence, Institute of Human Centred Engineering, Universiti Teknologi Malaysia, 81310 UTM Skudai, Johor, Malaysia.

* ajune@utm.my

Abstract. Learning is a process to extend the knowledge, but learning science helps to understand the world. Science education is important due to the relevance in human's lives and the universally applicable problem-solving and critical thinking skills it uses and develops. It is a foundation of education for student. When it comes to student, science education gives them the skills and knowledge that are required for their problem solving and critical skills. Science is the best subject that can be learn by experimenting and experiencing. Nowadays, technology had taken part in many domains. Same goes to education, technology had contributed many improvements to make the learning process more attractive. Hence, in this modern era, the implementation of video games in education can be the effective way for student to learn while having fun in learning. Therefore, this project aims to develop a science game for student to play while experiencing the new method of learning process. The game is developed by using Unity3D game engine.

1. Introduction

Educational game can be applied in many subjects taught at school. For this project, science subject for student Standard 1 is chosen. The game created can be a platform of new learning method since the subject requires student to explore their environment and nature. Hence, with the game version, students are able to play and learn science interactively without neglect the education elements. The content of the game is appropriate for the Malaysian education system which following the syllabus from new curriculum which is Kurikulum Standard Sekolah Rendah (KSSR).

Nowadays, there are many methods to improve the way of study in school. Over the years, educators had provided various way to help students gained a long-lasting interest in study because it has the potential to enhance the student's attention to stay focus and the learning quality [11]. Since the development has rapidly growing, the evolution of learning method changes in a modern way. One of the approaches to increase the quality of learning is by using a computer and games as a medium [9]. Game-based learning can be considered as the mix of education and games. Education game design must have a balance between fun and educational value [6]. Game can be said to be successful in help student to encourage problem solving and decision-making skills. It also helps to increase student understanding and time engagement in learning.

This paper contains six sections as follows: Section I presents the introduction of the paper. Next, Section II describes about the literature review that had been done for this project. Section III presents the methodology used for the project. The implementation of Science subject into the educational game is described in Chapter IV. Then, Section V described the test conducted and the evaluation of the result. Lastly, Section VI contains the conclusion of the project development.

2. Literature Review

2.1 Game and Education

Computer games have become immersed in our society and cultural environment. The advantages of computer games become the reason by many parties to use it as attraction especially in learning process. Game has unique perspectives of teaching and learning differ from any platform [10]. Computer game gives a new method and approach for educators to apply it in learning process. The games also have to be balance in entertainment and academic aspects. In addition, playing online games may be something which can enhance a child's learning and development. It also has potential in improving on the perceptual and motor skills [3]. In other research, there are four domains which of the positive effects in playing video games, i.e. cognitive, motivational, emotional, and social [2].

2.2 Game Based Learning

Game-based is an approach to learning while exploring the suitable aspects of games in a learning context. For example, teaching history with simulations. When the learn processed in groups, the finding shows that the students achieve better experience, learning performance and satisfaction [4]. There are many aspects of education which contributes to active learning. As stated by Pivec and Dondi, [9] the first aspect of educational role-play is the immersion and reasonable experiences in learning process. Other than that, the development of personal understanding about the topic that gains from the resources and lessons. Next, the personal method of understand the lesson with other person by testing and challenging each other knowledge. Lastly, the reflection of someone's achievement to put it as a self-benchmark. Based on the research, game-based learning enables the learners to understand better and helpful in improving the achievements and learning attitudes [13]. Green and Bavelier [3] stated that serious game motivates and trigger them to use their skills and ability in critical thinking. Collaborative learning environment has been studied by [13] and online teaching using mixed reality for telepresence [14].

2.3 Science Education

For this game development, the subject referred is Science for Grade 1. The syllabus applied following the latest curriculum in Malaysia education system which is Kurikulum Standard Sekolah Rendah (KSSR). The aim of the science education is to provide opportunities for pupils to learn about themselves and environment through experiencing and investigating [7]. In the other hand, it acquires knowledge and skills to enable pupils to apply it to make decisions and solve the problem in everyday life. As what had mentioned above, science requires more skills and knowledge by experimenting the things around them. Researcher stated that student's interest and willingness to study science is diminished since they require to rote [1]. Johnson and Mayer [5] review that digital game helpful in improving spatial cognition, perceptual-motor skills, problem solving skills and visual attention processing. Science is wide and many more can be discovered. The basic of science are taught since earlier of the age. Its knowledge can be gained through direct or indirect approaches. As usual, direct knowledge comes from the lesson at school or home. Meanwhile, indirect knowledge comes from the things happened in our surrounding. It also can be gained when something influenced and triggered us to solve the problem. In this project, the topic used in this subject for the game development is living things and non-living things.

2.4 Selected Topic

The selected topic chosen in this subject is from Chapter 3 which is Benda Hidup & Benda Bukan Hidup. There are four sub-sections under this chapter which are (1) Benda Hidup & Benda Bukan Hidup, (2) Ciri-ciri Benda Hidup, (3) Kecil dan Besar, and (4) Keperluan Asas Benda Hidup.

In overcoming the problem of quality consistency of science education, many experts called for inventing the approaches of science education. Game and simulation for science education are the suitable approach since it already a part of nowadays norm of community [1][7]. Furthermore, game provides better way in attracting student's attention and has different method in learning process. There are many digital games that can be obtained from internet or play store. Playing digital games not only limited in classroom but student also can play it at home as self-learning time. Below is some example of existing science games for kids. There are no existing games for Science Standard 1 from Malaysia. Hence, the example below shows the Science Games for Grade 1 from abroad schools which are (1) Science Game for Kids, (2) First Grade Learning Game and (3) Hermione 1st Grade Science Learning Education Game. Below is the table that describes the comparison of the existing games.

Table 1. Comparison of existing applications.

	Science Game for Kids	First Grade Learning Game	Hermione 1st Grade Science Learning Education Game
Operating System	Android, iOS	Android, iOS	iOS
Rating	3.0	4.8	1.0
Price	Free, additional fee to unlock other games	Free, additional fee to unlock other games	Fee required
Curiculum Based	Curriculum from worldwide syllabus	Curriculum from worldwide syllabus, includes another subject	Next Generation Science Standard (NGSS) Curriculum
View Based	2D	2D	2D
Features	-colourful image -attractive graphic -sound, music	-colourful image -attractive graphic -animation -sound, music	-colourful image -animation -sound, music

3. Methodology

This section explains the methodology used in this paper. Methodology is essentially a systematic procedures and act as a guide to ensure the development process is on the right track. There are four phases includes in the methodology to develop this research.

3.1 Phase 1 Preliminary investigation of the requirements for educational game

In this phase, the studies on Science subject for Standard 1 was conducted for the assigned topics. The studies were made to ensure the information is right based on the syllabus. Next, a research was made to investigate the effectiveness of the learning method through the game and the elements required to enhance the game appearance. The interview was conducted with the teachers from primary school and preschool to collect related information and idea for Science educational game development.

They had given their opinion about the application of game in learning method. Both interviewees are strongly agreed with the statement because the method can make the students more understand about what they had learned. Madam Noridayati had thought that the game can enlighten the learning environment and easy to stimulate the students thinking. Due to that, Madam Saodah shared that she had applied various physical game such as playing teng teng while counting. The students had given great response since they had enjoyable moments while learning with their friends. In addition, the teachers also agreed that the digital educational game can be a part of learning method. Madam Saodah shared that the digital game can attracts more attention of the student. Meanwhile, Madam Noridayati stated that students nowadays are more attracted to the gadget. Hence, the game can be one of the platforms for them to enhance their knowledge from what they had learned at the school.

3.2 Phase 2: Science Education Game Design

The game was designed based on the ideas and the requirements obtain from the previous phase. The game concept and game elements were designed to match with the suitability of the educational game.

To create a playable game, the game mechanism was designed based on the suitability for the target user. The game has three categories which represents the subtopics of the chosen topic which is living thing and non-living thing. The educational content of the game is based on the KSSR curriculum as discussed previously.

3.3 Phase 3: Science Education Game Design

They had given their opinion about the application of game in learning method. Both interviewees are strongly agreed with the statement because the method can make the students more understand about what they had learned. Madam Noridayati had thought that the game can enlighten the learning environment and easy to stimulate the students thinking. Due to that, Madam Saodah shared that she had applied various physical game such as playing teng teng while counting. The students had given great response since they had enjoyable moments while learning with their friends. In addition, the teachers also agreed that the digital educational game can be a part of learning method. Madam Saodah shared that the digital game can attracts more attention of the student. Meanwhile, Madam Noridayati stated that students nowadays are more attracted to the gadget. Hence, the game can be one of the platforms for them to enhance their knowledge from what they had learned at the school.

3.3 Phase 4: Development

To develop this project prototype, a suitable game engine was selected as it provides many functions and tools to create the Science educational game. Unity provides a tool to develop the games and interactive designs or experiences in 2D or 3D. For this game, the project was built on Windows PC platform so the players can play the game through their PC. To make the game functional, there are several game mechanics used in this game such as drag and drop, randomization, move player, timer, and scoring.

3.4 Phase 5: User Acceptance and Usability Testing

Usability Testing is a useful testing methodology to test and evaluate the usability aspects of an application. It is vital in improving the quality of user experience as usability testing emphasizes aspects such as ease of use, efficiency, and satisfaction, based on the research done by [15].

4. Implementation

This section explains the implementation of this paper based on the methodology of this paper.

4.1 Game Concept

The concept of the Science educational game for Standard 1 is to develop a 2D game based on the Science subject for the subtopic in Chapter 3; 'Living Things and Non-Living Things'. The game is a computer-based game which has three mini games indicates the subtopics. The first subtopic named living-things and non-living things. Next subtopic is the characteristics and basic need of living things and the last subtopic is small and big. Players are freely to choose which game that they want to play. The target audience of this educational mobile game is student in Standard 1.

4.2 Game Mechanics Design

In this game, there are three categories of game according to the subtopics of the Science subject. Each game has three levels and the difficulty is increase by the level. The player needs to follow the instruction given before they can start playing the game. In the first game, players are forwarded to the next level once they succeed with current level. Players are able to gain the award or feedback once the game is done. The total marks of the players were displayed too once the game is over.

The same thing goes to the second game and the third game. Each game has three level and the players proceed to the next level when the current level is done. In the second game, players must collect the basic need items as much as they can to get the highest marks. Each collected item contributes to 10 marks. The players must control the character's movement by using left and right key. The game is done when the player hits the non-basic need item. Once the game stop, the marks will be displayed, and the players can choose whether to replay the game or return to the main menu. The difficulties of the game increases by adding the speed of the falling object, so that the player need to take the fast action to avoid the unnecessary item.

Lastly, the third game requires player to drag the objects and sort them according to the size. The boxes are provided with the number listed. However, there is a time limit to complete the task. Hence, the players need to complete the task within the time given. As the level increases, the time limit is also getting shorter to increase the difficulty. The time taken will be recorded once the player succeeds the task and the players also can replay the game if they are failed to do so. The concept of the Science educational game for Standard 1 is to develop.

4.4 Game Level

Each mini game has different type of difficulties. For example, in the first game, the players need to drag the objects based on category of living things and non-living things. As the level increases, the objects are varied and changed. Next, in the second game, the players need to collect the basic need item that are falling randomly. They must avoid the non-basic need item to stay survive. When the level increase, the time duration between the two objects to fall are short, so the player needs to response quickly as the falling objects are increased. Lastly, for the third game, players need to succeed the task within the time given. As the level increases, the limit time given is shorter. Hence, players need to be quick in order to proceed to the next level.

5. Graphical User Interface

A properly designed Graphical User Interface (GUI) of the game enable the player to start the game easily.

Start page has two buttons to start the game or to quit the game. Main page has four buttons, while Login page allows user to enter their name. After the user logins to the game, it has three options. When user picks Information button on the Main page it will give the information on how to play the game and user also allowed to get into game setting. The game is able to on or off the audio. While the Dialogue System appears to check whether the user has learnt from the game especially about Chapter 3; 'Living Things and Non-Living Things'. There are three stages the user needs to complete. Figure 1 (a) shows the scoring board and Figure 1 (b) shows the Panel Note. It was for the user to do reading.

(a) (b)

Figure 1. (a) Scoring board (b) Panel Note

A list of interfaces include in this project are listed in Table 2 below.

Table 2. User Interfaces with Game Elements.

Game Elements and User Interface	
Start Page	Main Page

Login Page	Game Option
	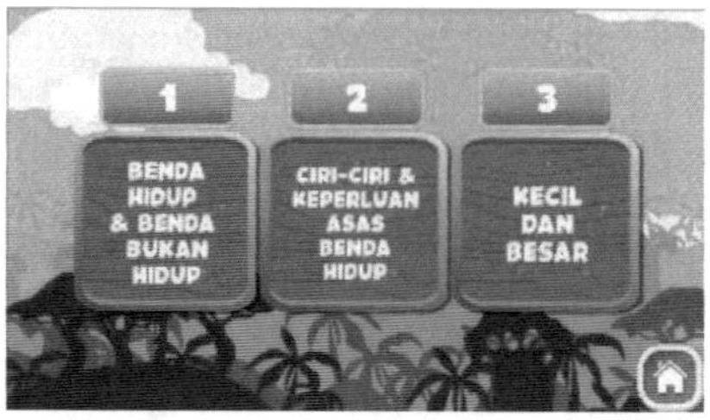
Information page	Music Setting

Dialogue System	First Game

Second game	Third Game
	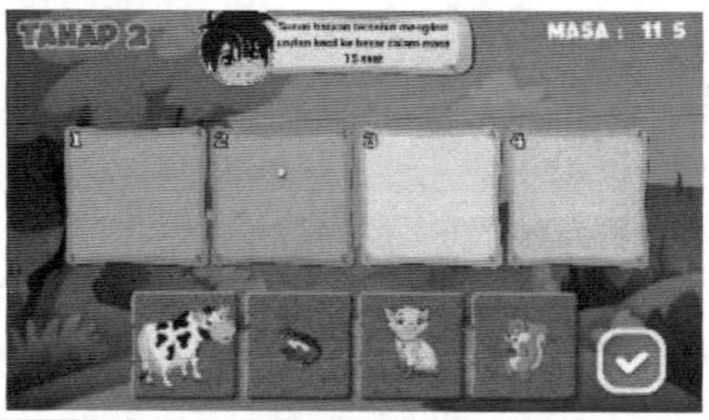

The implementation of code in game development is important to build the logic of the game. There were several classes implemented to ensure the game was functioning well. The game mechanics used are explained below.

5.1 Drag and Drop Interaction
Each mini game has different type of difficulties. For example, in the first game, the players need to drag the objects based on category of living things and non-living things. As the level increases, the objects are varied and changed. Next, in the second game, the players need to collect the basic need item The handler used was IPointerDownHandler, IBeginDragHandler, IEndDragHandler and IDragHandler. To detect the mouse clicks until it was released, the method used was OnPointerDown. The function applied when the mouse clicked on the object. Next, IBeginDragHandler and IEndDragHandler were used to detect when the objects are started to be dragged. When the player starts dragging the object, the OnBeginDrag method called to unblock the ray cast on the canvas. Otherwise, the ray cast is blocked once the player ended the drag using OnEndDrag method. Then, OnDrag function was called to enable the dragging objects move around the canvas. Meanwhile, in Drop class, the handler used was IDropHandler. The method called was OnDrop.

5.2 Randomize and Timer
In creating the questions for the game, it must generate randomly every time the player starts the game. In the first and the third game, the objects had been created to spawned randomly by switching its place or object. In second game, the objects are spawned randomly and falling within the assigned duration. Each object assigned with box collider and rigid body components. Box collider used to detect the hit between the object and rigid body used to apply the mass on the object.

In Kecil & Besar game, timer was applied to every level. Player has to complete the task within the required time. The game is over once the timer stopped. The timer applied in Update function to enable the timer decrease whenever the frame looped. Time.deltaTime was added to round the time per second. The time required was declared first according to the level. Timestart variable hold the timer required in the gameplay.

5.3 Movement and Scoring
Moving the player provided in the second game which is Ciri-ciri & Keperluan Asas Benda Hidup game. The inspector of the player was applied with Rigidbody 2D which enable the character to move horizontally in XY plane rotate on perpendicular axis. Vector2 was used to represents velocity of the object. The input received only required horizontal movement which is left and right.

Each correct answer was rewarded with 10 points. The score triggered when the objects were collided with the same tags. In first game, the dragged objects and dropped objects were declared based on the living things and non-living things categories. Meanwhile, in second game, the character was tagged with "need" to allow the scoring added when it is collided with the basic need objects

6. Evaluation and Results

6.1 User Acceptance Testing
The test was done as to determine whether the application can be accepted or not. The evaluation was done using Blackbox method. The feedback of the result was compared between the expected outcome and actual outcome. The testing was performed on the 10 respondents from age range between 7 to 10 years old. The respondents were selected randomly from the nearest community. The flexibility in selecting respondents from different age ranges was due to the COVID-19 pandemic which limiting the testing process. There are 3 target users aged between 10-12 years old and 10 non-target users aged between 23-30 years old are involved in user acceptance testing and usability testing. The table below shows the result of Beta testing using Blackbox method.

Table 4 shows the result of Beta testing by using Blackbox method. From the result, the events done by the respondents were mostly were successfully functioning. All the buttons that available in the game were work well and can be switched to the assigned scenes. The respondents also can view the score after they had done playing the game. The game mechanics such as drag drop and player movement are functioning well instead of a bit problem about the inertia of the character's movement. However, there were some problem in score when the respondents played the game. The score sometimes may be correct, and it can be negative value when the respondents were wrong at the beginning of the game.

The error might be from the mistakes in code itself. But overall, the test give positive feedback from the respondents as the flow of the game is works well and working as expected.

Table 4. User Interfaces with Game Elements.

Event	Expected Outcome	Actual Outcome
The users click on the setting menu	The setting menu is displayed	Can easily displayed setting menu
The users enter the dialogue menu	The dialogue menu displayed with button to proceed the dialogue	The dialogues are run smoothly and easily to proceed
The users click on the home button	The home interface displayed	Can easily displayed home menu
The players win or lose the game	The feedback panel displayed with score, home button and restart button	The feedback panel displayed with home button, restart button but the score sometimes displayed wrongly
The score added and deducted	The score added and deducted correctly	The score added correctly but deducting can be negative value
The users click on the audio button for mute/unmute	The audio is mute / unmute	The audio mute and unmute only works in home scene.
The users click on the pause button	The pause panel displayed with home button and resume button	The pause panel displayed with home button and resume button
The users click the items to drag and drop	The drag and drop activities are functioning	Can easily drag and drop the items
The users click on left and right keys to move the character	The character can move left and right	Can easily moves the character but the inertia is little bit over.

6.2 Usability Testing

Usability testing was done as to get the feedback from the respondents about their experience and satisfaction. The questionnaire contains two sections to be answered which are the first section required the information about the respondent and the second section provides post-test related questions. The post-test related questions are categorized into several factors such as efficiency, satisfaction, and learnability.

The first section in the questionnaire was asked about respondent's information such as gender and age. Besides, the experience in learning the topic and the experience in playing digital game using PC also had been asked. The test done on 10 respondents which consists of 5 males and 5 females from different age range. There were 4 respondents aged 7 years' old which is in Standard 1. Meanwhile, there were 1 respondent aged 8, 3 respondents age 9 and 2 respondents age 10. The respondents were selected in different age range due to the limitation during Restrictions Movement Order (RMO). Most of the respondents had learned the selected topic during Standard 1 and familiar with it. In the other hand, 6 respondents stated that they have no experience in playing digital game using PC while 4 respondents had experienced it.

From the results of non-target users as shown in Figure 2, most of the non-target strongly agreed that this game prototype was satisfying to use as 60% of non-target users choose "Strongly Agree" and 40% of non-target users choose "Agree" for the question related to the satisfaction. The non-target users felt satisfied to use the game prototype to learn English words. For the memorability of the game prototype,

IOP Conf. Series: Materials Science and Engineering **979** (2020) 012012 doi:10.1088/1757-899X/979/1/012012

most of the non-target users able to understand the function of interface elements such as buttons and input controls as well as gameplay of mission at each level of the educational mobile game prototype.

(a) Efficiency component

There are three questions asked to determine the effectiveness of the game flow when the respondents play the game. It is also determined their understanding about the system.

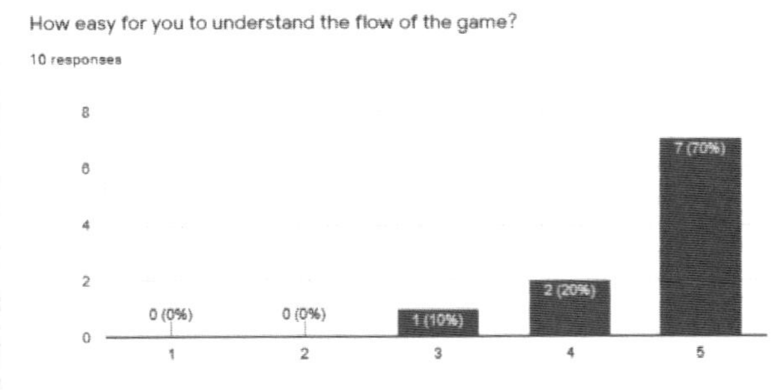

(a)

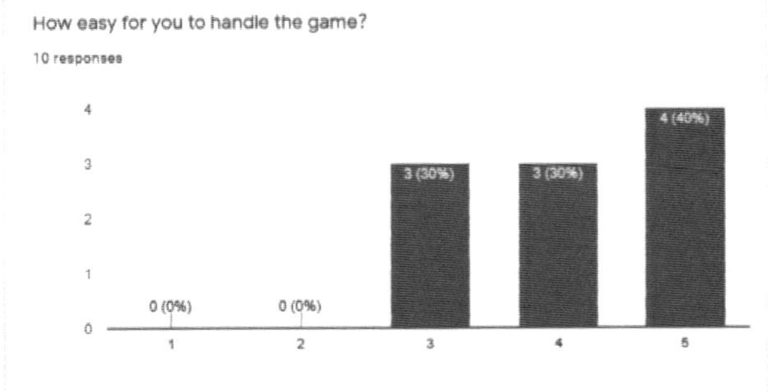

(b)

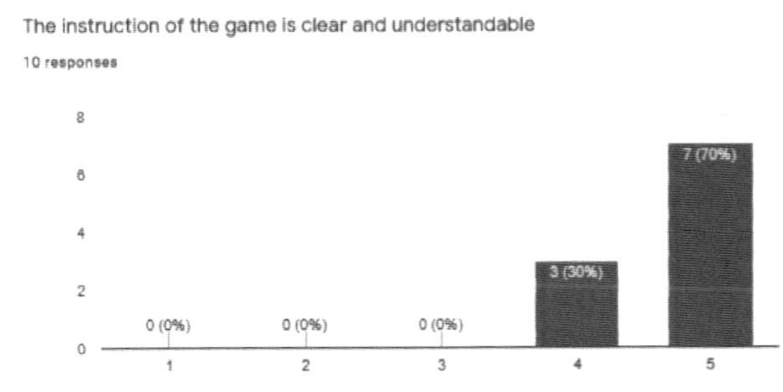

(c)

Figure 2. Efficiency elements

The result from the tests above shows different outcomes based on the respondent's understanding about the game flow. Figure 2 (a) illustrates the bar chart result of the respondent's understanding on the game flow from the scale 'very hard' to 'very easy'. There was 70% of respondents scaled that the game flow was easy to understand. However, one respondent scaled neutral because the lack of skill in playing game. Figure 2 (b) illustrates the graph about the respondent's skill in handling the game. The scale from 'very easy' to 'neutral' were chosen averagely because most of the respondents never experienced in playing the game using PC. Most of them were good in control the game but not efficiently as expected. Figure 2 (c) shows the result of how the respondent's feedback on the game instruction. 70%

of them were strongly agreed that the instruction was clear and understandable, and the rest was agreed with the statement.

(b) Satisfaction component

There were two questions asked about the respondent's feedback on the game design including game theme, fonts, colors, and interfaces together with sound effect and music background.

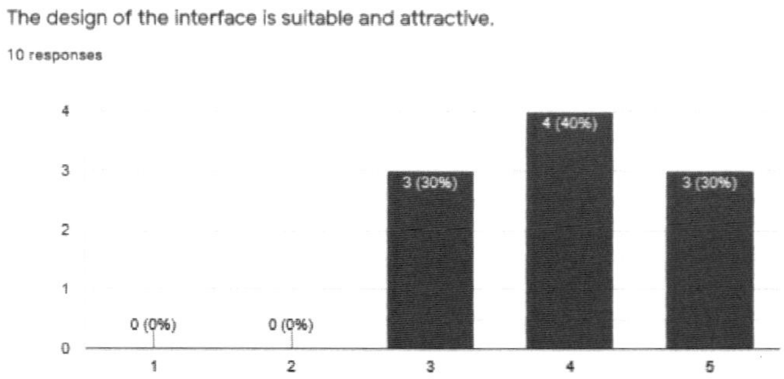

(a)

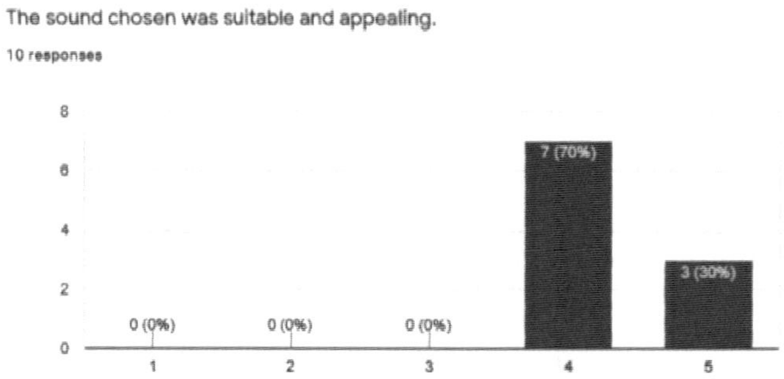

(b)

Figure 3. Game design factor questionnaire

Figure 3 (a) illustrates the graph of the feedback of the respondents regarding to the design of the game application. From the figure, most of them agree that the design of the interface was suitable and attractive. Meanwhile, the number of respondents that chose 'strongly agree' and 'neutral' was equal which is 30%. The suitability of the design was considered based on the age of the user and the type of the game which was educational game. In the other hand, Figure 3 (b) demonstrates the result of the feedback for the audio used in the game. The audio includes sound effect and music background. From the result, 3 respondents were strongly agreed that the sound was suitable and appealing. Meanwhile, the rest of the respondents were agreed to the statements. Overall, the audio used in the game was attractive and suitable according to the content of the game.

(c) Learnability component

There are three questions asked about the respondent's feedback on the game effect on them while playing the game. The effects include the respondent's reaction and respondent's knowledge.

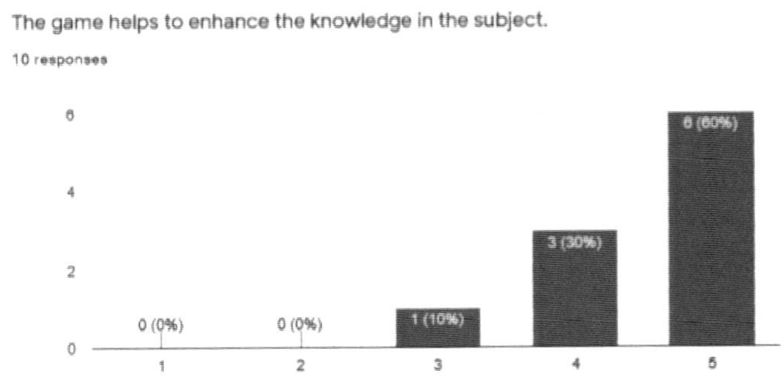

(a)

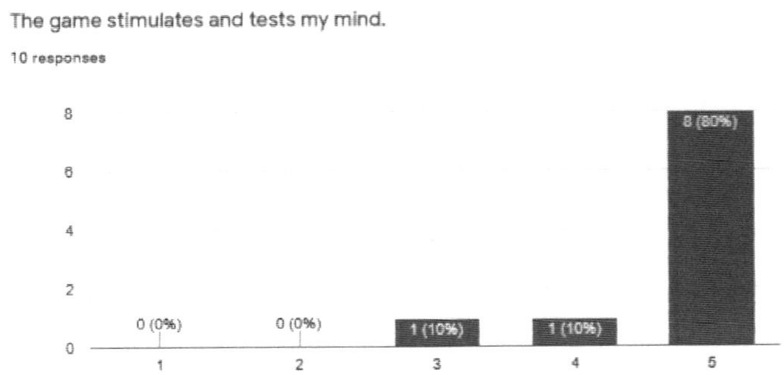

(b)

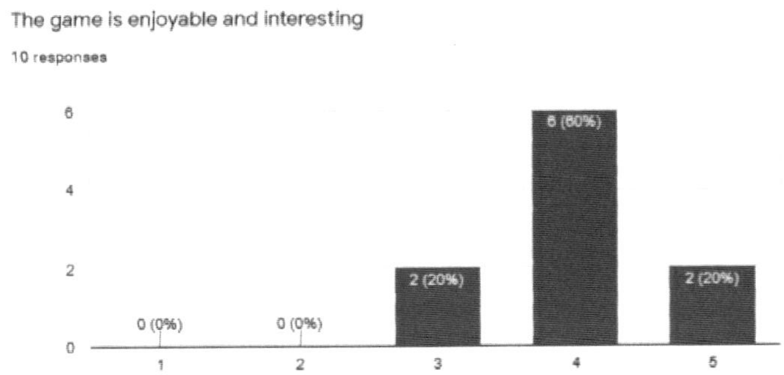

(c)

Figure 4. Game effect factor questionnaire

The figures above illustrate the results of the testing for the learnability on the respondents in terms of respondent's reaction and knowledge. Figure 4 (a) asked about the opinion of the respondent regarding to the statement as stated. 6 respondents were strongly agreed that the game helped them to enhance their knowledge on that topic. 3 of them were agreed and 1 respondent just feel neutral. This is due to the age of the respondent that was already 3 years older than the target respondent (7 years old).

Next, Figure 4 (b) asked the respondent's opinion about the game that can stimulates and tests their mind. From the result, it can be said that most of the respondents were strongly agreed that the game had stimulated and tested their mind. Meanwhile, 1 respondent was respectively agreed and feel neutral to the statement. That effect on the respondents was because of they must think how to strive the missions in the game. They had to revise what had they learned in the class while playing the mission in the game. The last question asked to the respondents is whether the game was enjoyable and interesting. 60% of the respondents agreed to the statement that the game was enjoyable and interesting while 20% respondent respectively agreed and feel neutral to the statement. Overall, the respondent's feedback is good, and they had felt satisfied playing this game

Acknowledgments
The authors would like to express sincere appreciation to all people involved in this project and contributed to the success of this project. We would like to convey our gratitude to Matching Grant (Q.J130000.3008.01M88) and Research Grant (Q.J130000.7308.4B429) for the funding.

References

[1] Council, N. R. (2011). Learning Science Through Computer Games and Simulations. Washington, DC: The National Academies Press.
[2] Granic, I., Lobel, A., and Engels, R. C. M. E. (2014). The benefits of playing video games. American Psychologist, 69(1), 66-78.
[3] Green, C. S., and Bavelier, D. (2003). Action video game modifies visual selective attention. Nature, 423(6939), 534-537.
[4] Hung, C.-Y., Sun, J. C.-Y., and Yu, P.-T. (2015). The benefits of a challenge: student motivation and flow experience in tablet-PC-game-based learning. Interactive Learning Environments, 23(2), 172-190.
[5] Johnson, C. I., and Mayer, R. E. (2010). Applying the self-explanation principle to multimedia learning in a computer-based game-like environment. Computers in Human Behavior, 26(6), 1246-1252.
[6] Moreno-Ger, P., Burgos, D., Martínez-Ortiz, I., Sierra, J. L., and Fernández-Manjón, B. (2008). Educational game design for online education. Computers in Human Behavior, 24(6), 2530-2540.
[7] Osborne, R., and Freyberg, P. (1985). Learning in Science. The implications of children's science. 70 Court Street, Portsmouth, NH: N/A.
[8] Pivec, M., and Dondi, C. (2004). Guidelines for Game-based Learning: Pabst Science Publ.
[9] Prensky, M. (2003). Digital game-based learning. Comput. Entertain., 1(1), 21-21.
[10] Squire, K., and Jenkins, H. (2011). Video Games and Learning: Teaching and Participatory Culture in the Digital Age: Teachers College Press.
[11] Yang, L.-H. (2010). Toward A Deeper Understanding of Student Interest or Lack of Interest in Science. Journal of College Science Teaching, 39(4), 10.
[12] Yien, Jui-Mei, H., Chun-Ming, H., Gwo-Jen, L., and Yueh-Chiao. (2011). A Game-Based Learning Approach to Improving Students' Learning Achievements in A Nutrition Course. Turkish Online Journal of Educational Technology, 10(2), 10.
[13] Ismail A W 2019 A review of the collaborative learning environment across virtual and augmented reality technology *In IOP Conf. Series: Materials Science and Engineering* (Vol 551, No 1, p 012050)
[14] Fadzli F E, Ismail A W, Aladin M Y F and Othman N Z S 2020 A Review of Mixed Reality Telepresence. *In IOP Conf. Series: Materials Science and Engineering* (Vol 864, No 1, p 012081)

IOP Conf. Series: Materials Science and Engineering **979** (2020) 012013 doi:10.1088/1757-899X/979/1/012013

Augmented reality battleship board game with holographic display

F E Fadzli [1,2] *, A W Ismail[1,2], M F A Rosman[1,2], N M Suaib [1,2], M S M Rahim[2] and I Ismail [3]

[1] Mixed and Virtual Environment Research Lab (mivielab), ViCubeLab, Universiti Teknologi Malaysia, 81310 Johor, Malaysia
[2] School of Computing, Faculty of Engineering, Universiti Teknologi Malaysia, 81310 Johor, Malaysia
[3] Faculty of Informatics & Computing, Universiti Sultan Zainal Abidin, Kampung Gong Badak, 21300, Terengganu

* fazliaty.edora@gmail.com

Abstract. The board game is a game that played on the board, and the Battleship board game is one of the examples. While developing the virtual board game, multi-user interaction is essential as a board game played by more than one people. To enhance the user experience, the board game has been implemented with Augmented Reality (AR) technology. AR can superimpose it into the real-world since it has the capabilities to overlay the virtual game on the top of the real world. However, AR handheld has limited perspective viewing for the viewers to see the battleship game. Therefore, this project aims to develop an AR battleship board game with a holographic display. Three phases have been carried out, first is to study the multi-user interaction for the AR battleship board game and holographic display with a different perspective. Second is to develop the multi-user interaction for AR Battleship board game. Finally, the third phase is to integrate the AR Battleship board game with a pyramid holographic display. The evaluation has been performed, and the result is based on the usability and user acceptance have been recorded. Based on the result, this project has successfully produced AR Battleship board game with holographic display application and significantly enhance the user experience.

1. Introduction

The board game is generally any games of strategy or chance played on a specially designed board and involved two or more opponents moving pieces, and using dice or cards [1]. Some examples of the board game are checkers, chess, snake and ladder, and monopoly as the board games involve more than one people, players able to explore and interrogate the knowledge in a fun and interactive way by working in groups or provoking the players [1]. This paper describes the implementation of the multi-user interaction of the board game by using the Photon Unity Networking (PUN). Augmented Reality (AR) is a combination of the real and virtual object in a real environment, interactively run and in real-time, and registered or aligned real and virtual objects with each other [2]. AR tracking is the main components in application development. One of the tracking methods applied is marker-based tracking. The AR can serve to display three-dimensional (3D) objects and their information when scanning the marker [3]. To implement the AR marker-based tracking, Vuforia SDK has been implemented during the application development. The implementation of AR provides a gaming experience for the player. The developed

IOP Conf. Series: Materials Science and Engineering **979** (2020) 012013 doi:10.1088/1757-899X/979/1/012013

board game had integrated with the holographic display to give the gaming experience for the non-player. The hologram is an image that can appear or display in the form of 3D and can be seen by naked eyes. The holographic display becomes more enjoyable when multiple points of views of the display shared by multiple users at the same time [4]. The sharing process can occur when using the four-sided hologram pyramid based on Pepper's Ghost Illusion concept.

This paper explains the phases to develop AR board game with a holographic display. This work needs to produce game elements for battleship and multi-player algorithm to execute the battle. Therefore, this paper discusses the proposed multi-user interaction before implementing the holographic setup for non-users. This research needs to obtain non-user experience if the battleship game can be displayed to see by non-users openly. While the multi-user interaction also collected to know the user experience with the AR interface.

2. AR Board Game

People already familiar with the AR during this modern era. It is a technology that allows the users to have an interactive experience with digital or computer-generated information and overlays over the real-world environment. That digital information can be either images, audio, video or gesture [5]. Kipper and Rampolla [5] stated in their book that AR is the combination of many types of technologies that bring digital information into visual perception. The augmented perception of reality and the creation of an artificial environment are those functions of the AR. Based on the book written [5], AR act as a tool designed to enhance the surrounding environment from the perception of the user. They also claimed that in some cases, AR is not only for the entertainment process, but it is beneficial to assist users in the process of decision-making. This is a significant element when AR has used in a game, especially a battleship game which required decision-making [6]. Figure 1 shows the AR battleship game.

(a) Real-world (b) AR Battleship

Figure 1. AR Battleship [6].

Battleship consists of four grid board, and the interaction was the ballistic shots. There was an interactive cursor based on the grid cell definition; it allows a visual hit. The application is defined in [6] by using ARToolkit tracking system which black and white marker was used as in figure 1 (a), while figure 1 (b) shows the whole game scene appeared on top of the marker.

Besides ARToolkit, marker detection can be performed by Vuforia SDK, especially for mobile application [6]. Camera images are making use of marker tracking to find square optical markers and estimate their relative pose to the camera [7]. It was also able to superimpose the camera image and the virtual object by using the camera image as the background in their display, which is the real world and using the marker pose. The augmentation stays on the marker although either the marker or camera moving.

The board game is commonly any games of strategy or chance played on a unique design board and involved two or more opponents moving pieces and using dice or cards [1]. Some examples of the board game are checkers, chess, snake and ladder, and monopoly as the board games involve more than one people, players able to explore and interrogate the knowledge in fun and interactive way by working in groups or provoking the players [1]. While sitting around the table, board games can be played by several people [8]. Majority of the board game is the multi-player game. Multiplayer is the involvement of multiple players to play the same game and interact within the same game environment [9]. This approach is in contrast with the single-player.

Table 1. Differences between single-player and multi-player [9].

	Single-player	Spacing
Number of players	One	More than one
Form of social play	The player can play with a friend, but the game environment cannot interact	More than one player can interact with the game environment
Type of rule	Formal rule	Social rule
Data-intensive suitability	More	Less
Data-intensive outcome	More predictable and less diverse	Less predictable and more diverse
Type of interaction	Linear interaction	Nonlinear interaction
Present of pressure	No, unless played with a friend	Yes
Type of learning	Individual learning	Social learning

Table 1 shows the differences between single-player and multi-player. The result indicates that multi-player can involve more than one player than single player. The form of social play of multi-player is where more than one player can interact with the game environment, but for the single-player, the player can be interacted by a friend but not with the game environment. The rule applied in multi-player is the social rule but the formal rule for single player. Less data-intensive suitability for multi-player as the outcome is less predictable and more diverse, but those factors were opposite with single player. At the same time, playing with multi-player, the pressure and be present but not in single-player, unless they played with their friend. Multiplayer involved social learning but individual learning involved in single player.

Besides the board game, battleship game also can be classified as a location-based game [10]. Battleship game usually can be played by two players [11]. Each player has marked game board that looks like checkerboard square. The players had to select any one or more square on their board by marking, covering or designating the selected squares as each selection being called a "shot" for each turn. The objective of the game is to locate and sink the other player's ship. The selected panel squares are corresponding with other player's panel squares. The ship is usually consisting of the straight line of square arranged, which can be either horizontal, vertical, or diagonally on the plane. The game concept remains the same during 2019. Each player must place their ship secretly on a 10-by-10 grid [12]. Based on the article written by [12], the grid was noted by a row and column indicator. Each player has two grids, and those grids were one with their ship and the other used for their opponent's grid hits record. By placing missiles, which is small pegs on the chosen location on the grid or cells, each player takes a turn to hit and sink their opponent ships. The first player, who managed to sink all the opponent's ships become the winner.

3. Holographic Display

3D image reproduced from interference pattern by a coherent light beam is the definition of hologram [13]. When a point source of the light, which is the reference beam of fixed wavelength encounters light of the same fixed wavelength arriving from an object, which is object beam, recording of the hologram in a 2D or 3D medium of the interference pattern is formed. Based on the article written by Ratan & Akshay [13], holography is the making holograms science which was usually intended for displaying

3D objects or can be simplified as a process of reconstructing and writing a hologram. Holography also a physical structure that diffracted the light source into an image. By shining a laser through a hologram and projecting an image on the screen, the holographic image can be seen looking into an illuminated holographic print. Some of the technique to display the holographic image is using an illusion of hologram on the screen, such as Pepper's Ghost.

The holographic display became enjoyable if the display can be shared by multiple visitors at the same time exploiting its multiple points of view [4]. Based on the article written by [4], multiscreen projection especially when using the pyramid glass allows visualization of a projection that proves to be just like a hologram, and it can view from different viewpoint simultaneously. Visualization pyramid was designed as the reflective materials that reflect the light from the holographic projection system on the monitor to display the 'floating' object. Holographic can be displayed using the four-sided pyramid. The method also can produce a 3D object as the holographic result in the real-world environment [14]. Moreover, the method also allowed the user to see the view from different angles and different perspective of the hologram. The object can be displayed from four distorted views, which are front, back, left and right and it is displayed as a floating object in the middle of the pyramid when implementing it with four-sided faces. While developing it in Unity 3D, four cameras used to render the texture facet and placed at four viewports of the object, which are front, back, left, and right view to project on each side of the pyramid. Besides that, four reflective side holographic pyramid and a monitor used as the main hardware for this holographic setup. The monitor displayed the texture facet, and the image reflected at the 45 degrees angle on the reflective surface.

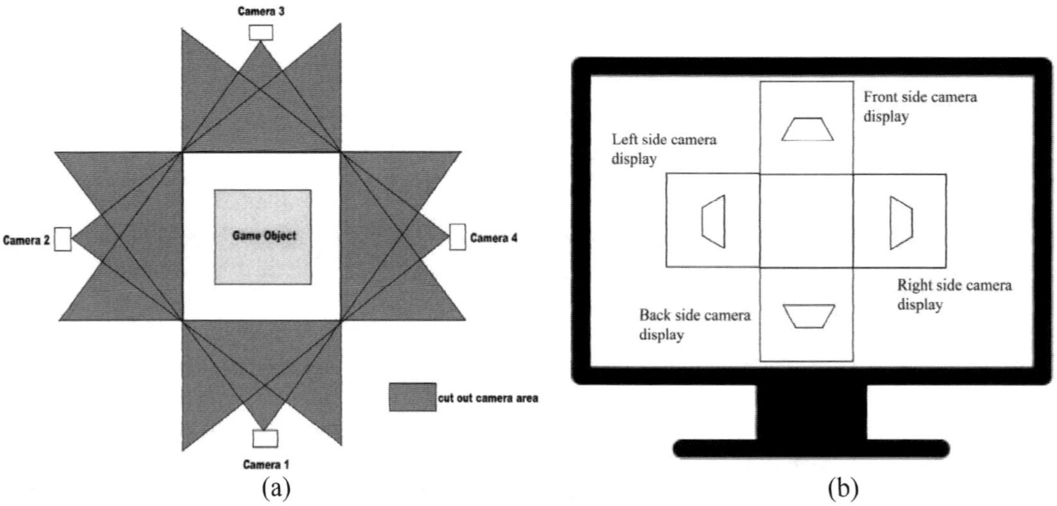

(a) (b)

Figure 2. Holographic setup.

The proposed holographic display in this project is based on Pepper's Ghost prototype. There are four cameras, which had used and placed at a different side of the game object. Those sides are front, back, left and right sides. Those cameras displayed those four views at the same time inside one desktop monitor to test the setup. Figure 2 (a) shows the camera placement during the holographic setup. The cut-out camera area indicates the area that had cut out from the camera view. This method is used to change the camera view, the changes from a perspective view to the parallel view. Figure 2 (b) shows the projection of the holographic on the desktop monitor.

The other component such as a holographic pyramid or pyramid glass also one of the essential parts for holographic display, as this method applied the Pepper's Ghost concept. The pyramid then turned upside-down and placed at the centre of the previous holographic setup, on the desktop monitor. Each side of the pyramid must be facing straight to each screen that displayed on the monitor screen, respectively.

IOP Conf. Series: Materials Science and Engineering **979** (2020) 012013 doi:10.1088/1757-899X/979/1/012013

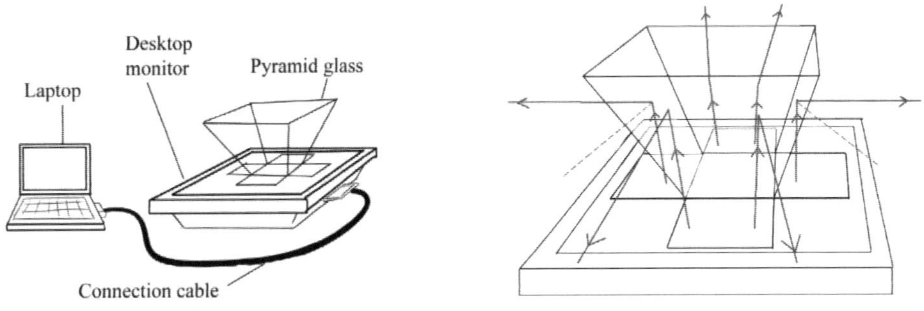

(a) (b)

Figure 3. The light ray from holographic projection reflected on the holographic pyramid.

The holographic pyramid displays with the laptop attached to the monitor are in Figure 3 (a). Based on Figure 3 (b), it shows that the light ray reflected on the pyramid glass from the holographic projection. The light ray from holographic projection (red line) hit the pyramid glass and then reflected toward the viewer (blue line). The angle of incidence and the reflection is equal, and the value of both angles depend on the normal line (cutting line) on the reflected faces.

4. Implementation

4.1. Game Elements

The board had made up by 6x6 or plane objects. Table 2 shows the battleship models for the battleship pieces; it was available as a free source on the internet.

Table 2. List of battleship models in the proposed battleship board game.

Kriegsmarine Schiffe (KMS) Bismarck	United State Ship (USS) Enterprise	Her Majesty's Ship (HMS) Belfast	Imperial Japanese Navy (IJN) Atago	Imperial Japanese Navy (IJN) Shimakaze

The interaction between the player and the game object had performed to apply the board game's gameplay. For example, when the player presses the directional arrow key as in Figure 4, the game object also moved the same as the selected direction.

- Controller: To control the ships or cursor left, right, up and down
- *Next* button: To place the ship and select the next ship
- *Rotate* button: To rotate the ships horizontally or vertically

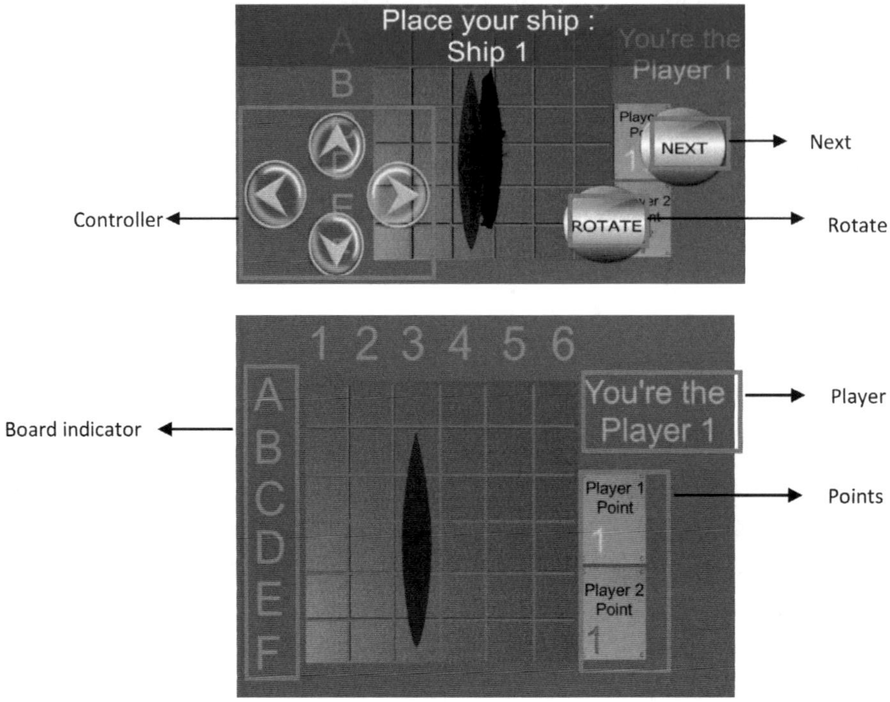

Figure 4. Controller and buttons in the battleship board game.

Controller: To control the ships Board and user feedback, as shown in Figure 5, there are three indicators:

(a) Board indicator: the cells on 6X6 board ([A, B, C, D, E, F] X [1, 2, 3, 4, 5, 6])
(b) Player: who was active in the game
(c) Player Point: the player 1's scoring
(d) Enemy Point: the player 2's scoring

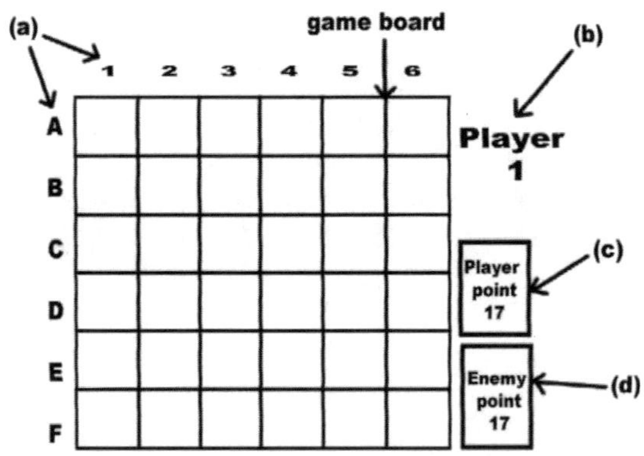

Figure 5. Board and feedbacks in the battleship board game.

The opponent's coordinate passed from the opponent to the player, and the Player 1 is indicated as an active player (in Figure 5). Figure 5 also shows the Points indicator when the player selects the target and the target same as the opponent's ship coordinate, the reduction of opponent point happened.

Battleship board game has its own game rules. The lists below some are the game rules that had followed:

- The ship must be placed within on the board only. The ship cannot move out of the board area. The ships only allowed placing horizontally and vertically. The ships cannot be replaced after the game starts
- The player that gets the turn must select the board square and say, "shot". Both players must mark the selected board square.
- If the opponent's selected board square has his/her battleship, the opponent must say, "hit" to tell that the target hit the target. Otherwise, the opponent says "miss" if the target hit the opponent.
- The winner is based on the player that manages to sink the entire opponent's ship first.

As the game involve multi-player, the social rule had applied, which mean that the gameplay can play variously, but the game concept has remained the same. The only changes that applied to the gameplay are for each turn, the player able to select three locations at the same time. After choosing the targets, the player ends the turn, and the opponent gets the turn.

4.2. Multi-players Interaction

As the developed Battleship board game played by two players, the game had integrated with multi-user interaction. Multi-user interaction is crucial to enable multi-players' interaction to allow the board game is playable for multi-player and allow the communication to occur between the players [15].

Game information had sent and retrieved between the players. Inside the same game room, there are only limited to two players at max. So, the interaction happened between both players; Figure 6 shows two players used two handheld devices to track a marker.

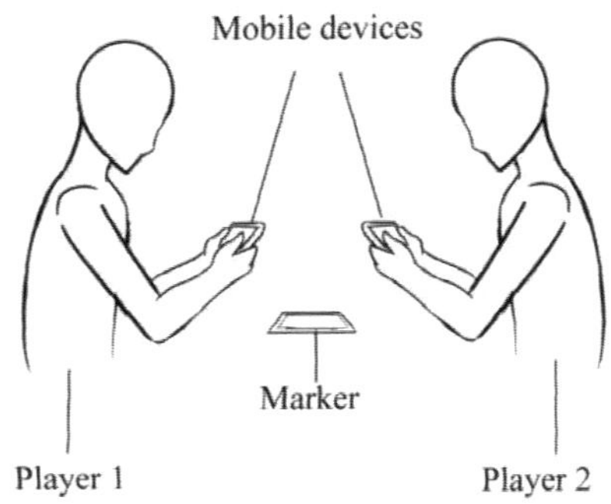

Figure 6. Two users in AR battleship board game.

Using networking to connect two players in the board game, the multi-player connection has been connected using internet protocol TCP, to send and retrieve the data in real-time. Implementing multi-user interaction, the Photon library is invoked into the game logic. A player was able to play with other players within the same room. The players had joined the room randomly. If there is a room available, the player joined the room but if no room available, the player has to create the new room. The player, who creates the room need to wait for any player to join the room before starting the match. The flowchart to implement this Photon Cloud service is in Figure 7. Based on Figure 7, when the player gets the turn, the player can interact with the application while the other player deactivates. After the player ends their turn, then the other player's turn has activated. Before the player creates the game or a room, they either can be a player or non-player. The holographic display or display hologram is setting up for non-players so they can see the battle.

When the player selects the board tiles, the tiles turned to blue, and the opponent tile turned to red. The opponent's ship coordinates shared with the player, so when the player selects the target and the target same as the opponent's ship coordinate. It affected the enemy point. When the player's ship had attacked, the player's point reduced and the point sent to the opponent.

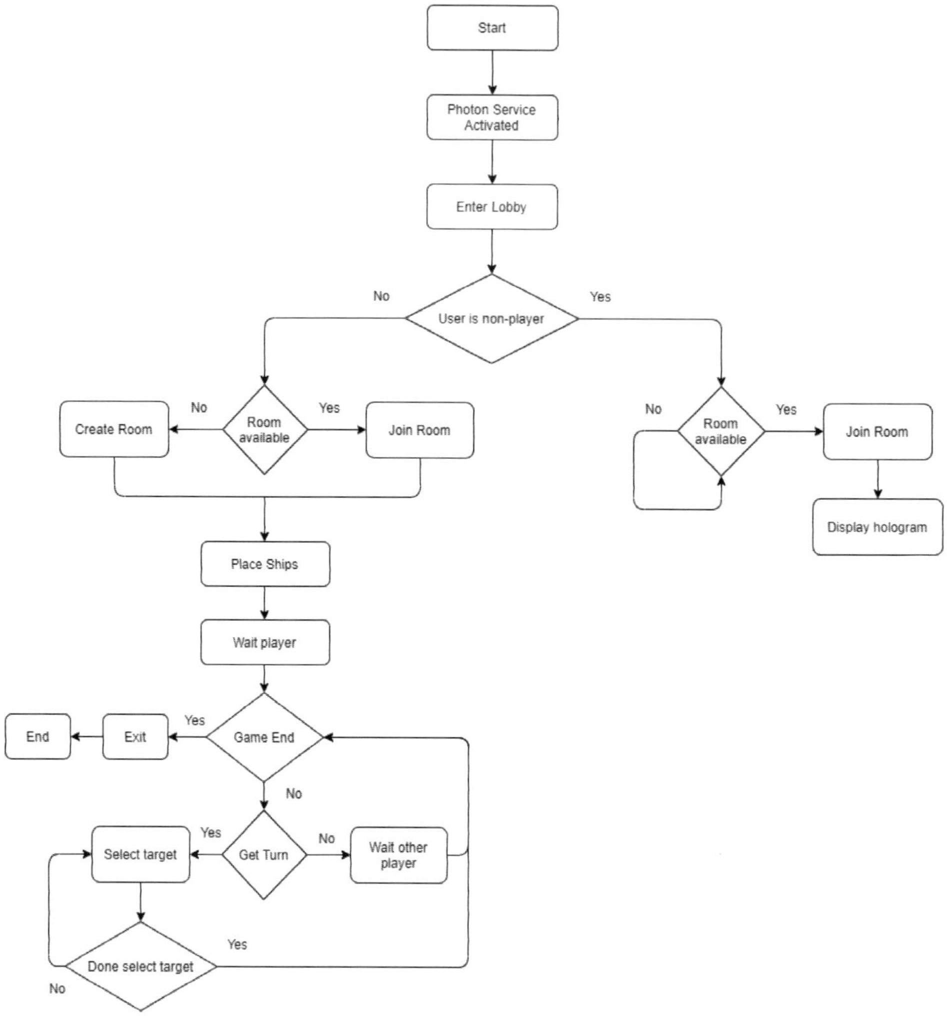

Figure 7. Battleship board game AR with holographic display flowchart.

4.3. AR Marker Tracking

Figure 8 shows before adding the AR tracking and creating the marker is when the tracker had registered. After registering the marker, the features on the image had produced. The process of converting the RGB image into greyscale, as illustrated in Figure 8, and the greyscale image has used to create the features. This process is known as feature-based tracking. The Battleship works with AR only after the handheld device captures the marker, and the application has recognized the feature to project the battleship board game on the top of the real-world (as in Figure 9). After integrating the Battleship board game with AR tracking, the AR board game has combined with a holographic display.

IOP Conf. Series: Materials Science and Engineering **979** (2020) 012013 doi:10.1088/1757-899X/979/1/012013

(a) RGB image was converting process into features

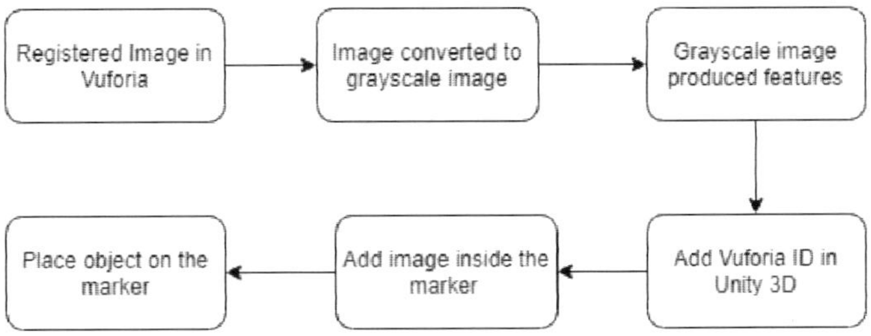

(b) Feature-based tracking process

Figure 8. Enabling AR for Battleship Board Game.

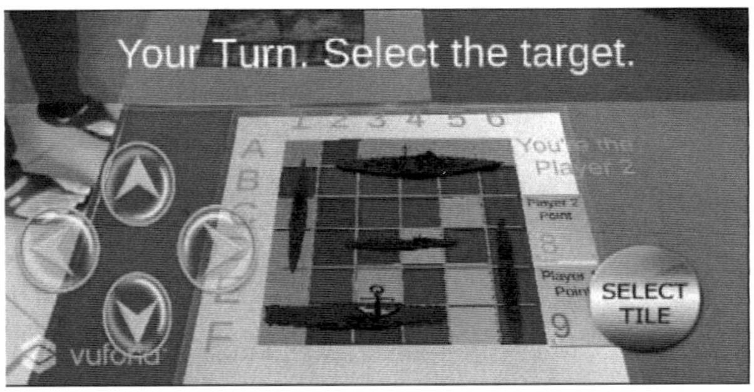

Figure 9. AR Battleship Board Game overlaid on the real-world.

5. Results

Figure 10 demonstrates the application required two players and at most four non-players. The physical markers have placed on the table, and the direction is facing the player (as in Figure 10 (a)). For the players, they used the mobile devices that have a built-in camera that used to track the AR marker while for the non-players, they used the desktop monitor, which connected to the laptop to display the hologram and pyramid glass placed on it. The 12-inch desktop monitor placed on the box, as shown in Figure 10 (b), and there is a cable connected horizontally. When the game has started, both players need to place the ships. After the first player was placing the ships, he needs to wait for the second player. After both players have finalized the ships, the battleship has initiated.

(a) AR Battleship Board Game with two players

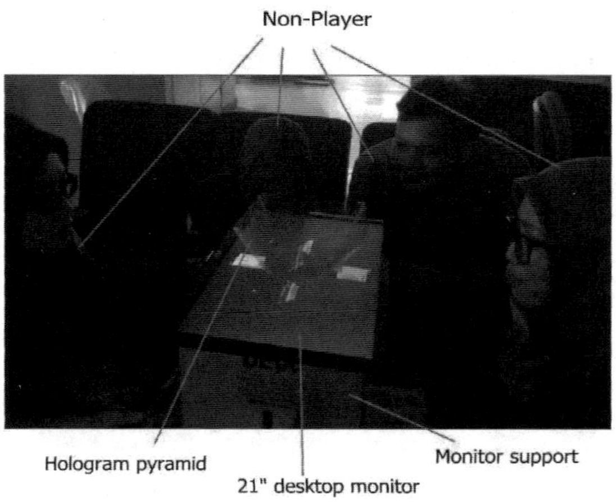

(b) Non-player with the holographic display

Figure 10. AR Battleship Board Game with users.

The player who gets the turn can select three tiles in one turn as shown in Figure 11 (a), while the second player needs to wait as shown in Figure 11 (b) until the first player's turn ended. The 'blue' colour of the tiles indicate that the player had selected the tile. The 'red' colour of the tiles indicate that the other player had selected the tile. The 'purple' colour of the tile indicates that both players had selected the tile. The player's point that reached 0 will be defeated, and the other player announced as the winner. For the non-player, they can watch the live game match through the holographic display. Only information about the ships position and changes of board tiles colour send from both players for holographic display. The location for players or non-players can be anywhere since the application has enabled internet connection and handheld devices can be portable for both players.

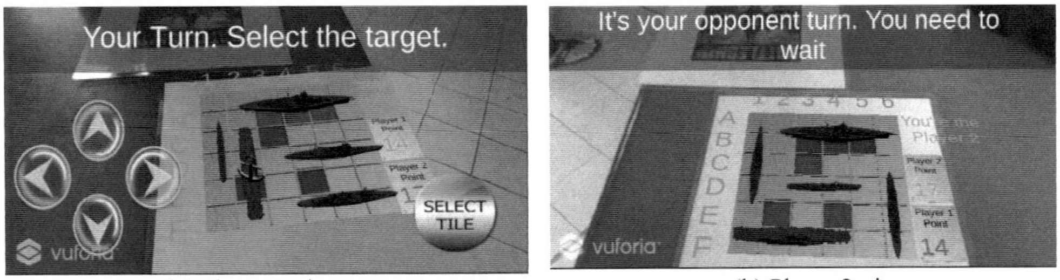

(a) Player 1 view (b) Player 2 view

Figure 11. AR Battleship Board Game in two handheld devices.

6. Conclusion

In conclusion, the Battleship board game has its own game goal. The goal is the player, who managed to sink all the opponent's ships will become the winner. AR Battleship board game with the holographic display is the outcome of the work. The board game has chosen for this work is the Battleship board game. Players can play the game, while for the non-player, they can watch the game through the holographic display. Battleship board game has implemented with AR and holographic technology, aims to enhance the user experience. However, handheld AR interface has limited perspective viewing for the viewers to see the battleship game.

Hence, this study aims to explore the AR battleship board game with a holographic display. Three phases have been carried out; the crucial is the implementation of multi-user interaction. Integrating the AR Battleship board game with a pyramid holographic display has provided the non-players with an alternative way to see the battleship. Future work is to evaluate the usability, to acquire the user experience results.

Acknowledgement

We would like to express our appreciation to Mixed and Virtual Reality Laboratory (mivielab), ViCubeLab at Universiti Teknologi Malaysia (UTM) for the facilities and technical supports. Special thanks to UTMHIR Grants (Q.J130000.2451.04G77) for the funding.

References

[1] Putri N, Setiyadi B and Nabila S 2018 The Implementation of Board Game to Improve Students' Speaking Achievement U-JET 7(2)

[2] Azuma R, Baillot Y, Behringer R, Feiner S, Julier S and MacIntyre B 2001 Recent advances in augmented reality *IEEE computer graphics and applications* 21(6), 34-47.

[3] Haryanto E V, Lubis E L, Saleh A and Lubis N I 2019 Implementation of Augmented Reality of Android Based Animal Recognition using Marker Based Tracking Methods *In Journal of Physics: Conf. Series* Vol 1361 No 1 p 012019

[4] Bovier F, Caggianese G, Pietro G D, Gallo L and Neroni P 2016 An Interactive 3D Holographic Pyramid for Museum Exhibition *In 2016 12th Int. Conf. on Signal-Image Technology & Internet-Based Systems (SITIS)* pp 428-34

[5] Kipper G and Rampolla J 2012 Augmented Reality: an emerging technologies guide to AR. *Elsevier*

[6] Silva D C and Vinhas V 2007 An Interactive Augmented Reality Battleship Game Implementation *Proc. Learning with Games* 213-219.

[7] Ćuković S, Gattullo M, Pankratz F, Devedžić G, Carrabba E and Baizid K 2015 Marker based vs. natural feature tracking augmented reality visualization of the 3D foot phantom *In Proc.of the Int. Conf. on Electrical and Bio-Medical Engineering, Clean Energy and Green Computing.*

[8] Burgun K 2012 Game design theory: A new philosophy for understanding games *CRC Press*

[9] Harteveld C and Bekebrede G 2011 Learning in single-versus multiplayer games: The more the merrier? *Simulation & Gaming* 42(1) 43-63.

[10] Bidwell N J and Holdsworth J 2006 Battleship by foot: learning by designing a mixed reality game *In Proc.of the 3rd Australasian conf. on Interactive entertainment* pp 67-74
[11] Ferreira A L S, Marciano J N, de Miranda L C and de Miranda E E C 2014 Understanding and proposing a design rationale of digital games based on brain-computer interface: Results of the admiralmind battleship study *SBC Journal on Interactive Systems* 5(1) 3-15.
[12] Autin M A and Gerstenschlager N E 2019 Battleship and the negative hypergeometric distribution *Teaching Statistics* 41(1), 3-7.
[13] Ratan A 2015 Holography - Working Principle and Applications 10.13140/RG.2.1.2688.4649.
[14] Siang C V, Isham M I M, Mohamed F, Yusoff Y A, Mokhtar M K, Tomi B and Selamat A 2017 Interactive holographic application using augmented reality EduCard and 3D holographic pyramid for interactive and immersive learning. *In 2017 IEEE Conf. on e-Learning, e-Management and e-Services (IC3e)* pp 73-78
[15] Nor' a M N A and Ismail, AW 2019, Integrating Virtual Reality and Augmented Reality in a Collaborative User Interface *International Journal of Innovative Computing* 9(2).

IOP Conf. Series: Materials Science and Engineering **979** (2020) 012014 doi:10.1088/1757-899X/979/1/012014

Design and development of learning mathematics game for primary school using handheld augmented reality

E Song[1*], N M Suaib[1,2], A J Sihes[3], R Alwee[1] and Z Mohd Yunos[1]

[1]School of Computing, Faculty of Engineering, Universiti Teknologi Malaysia, 81310 Johor, Malaysia
[2]UTM Big Data Center, Ibnu Sina Institute of Scientific and Industrial Research, Universiti Teknologi Malaysia, 81310 UTM Skudai, Johor, Malaysia.
[3]School of Education, Faculty of Social Sciences and Humanities, Universiti Teknologi Malaysia, 81310 UTM Skudai, Johor, Malaysia.

*elaine.song7@gmail.com

Abstract. The traditional learning methods have some drawbacks and limitation such as restricted learning space, high cost, and lack of student engagement. In the era of digitalization, traditional education starts to transform into a more dynamic and modern education with the influence of the 21st-century learning method. Moreover, due to the pandemic, schools and universities are moving to digital learning to contain the spread of Coronavirus Disease 2019 (COVID-19). Therefore, the aim of this project is to develop a Mathematics game using handheld Augmented Reality (HAR) for primary school children as an alternative method in learning Mathematics. There are three main phases have been carried out, first is to study the requirements of an educational game. The second phase involves the design and development including the game element, storyline, game characters as well as user interface design. Finally, the third phase is integrating the learning Mathematics game using HAR, followed by evaluation. The evaluation of the learning Mathematics game was carried out for usability and user acceptance. The usability results show that the users are satisfied with the prototype and feedbacks show the game is fun. Overall, this project has successfully produced a learning Mathematics game using HAR for primary school children.

1. Introduction

In the era of digitalization, traditional education starts to transform into a more dynamic and modern education with the influence of 21st-century learning method. A modern education system focuses mainly on using technology in education. One of the common uses of technology in education is digital game-based learning also known as an educational game. Today kids grow in a world full of the advancement of technology, a lot of electronic games that targeting our youth. This causes kids today to spend leisure time playing electronic games is not an extraordinary thing [1]. Besides, there are many researchers argued that electronic game such as the educational game is potential to support learning and it shows a positive improvement in various domains such as science, statistics, math, and biology [2].

Recent year Augmented Reality [2] has evolved as a popular technology and rapidly been used in the fields of business, gaming, logistics and others [3]. There are various studies have shown AR technology offers many advantages in the education field. For example, a digital immersion provides by AR to build a student's confidence in solving problems in the real world [4]. This paper will discuss about a proposed game focusing on learning Mathematics that combines the use of handheld AR technology and it is specifically designed for primary school children to strengthen the knowledge on the topic fractions. The outline of this paper will be as follows; Section 2 presents the related literature review, Section 3 outlines the methodology and Section 4 explains the implementation of this project. Sections 5 and 6 respectively discuss on the results and testing, while Section 7 summarizes the paper.

2. Literature Review

2.1. Background of Mathematics

Mathematics was defined as the abstract science of quantity, number, and space either as abstract concepts whereas applied Mathematics was defined as Mathematics that is applied to other disciplines, for example, physics and engineering [5]. Mathematics is not only one of the compulsory subjects in Malaysia's education system and it is also one of Science, Technology, Engineering and Mathematics (STEM) subject. According to the Education Blueprint 2013-2025 published by Ministry of Education Malaysia, the aims to implement STEM education is to prepare students with the skills to meet the science and technology challenges and to ensure that Malaysia has a sufficient number of qualified STEM graduates. In primary school, children learn about subtopics like the whole number, fractions, decimals, percentage and money in the topic of number and operation. This proposed project focused on the subtopic of fractions.

2.1.1. Fraction

Fractions define as a numerical quantity that is not a whole number [5]. A fraction is telling a "part" of a "whole" and the number is between zero to one. The number on top of the horizontal line call numerator and the lower part of the number call denominator. Figure 1 shows the fraction 2/3. 2/3 represented the shaded areas and it means that the whole is divided into 3 equal parts, 2 parts of the same is considered. Therefore, 2 is the numerator while 3 is the denominator. It can read as 2 over 3.

Figure 1. Fraction of 2/3.

2.1.2. Mistake and Error in Learning Fraction

There are a lot of researches and scholars from different education paradigms begun to identify the main problems encountered by the primary school students in learning the topic of fraction and provide an explanation about the problems. The common errors and possible common mistakes did by the student were identified by Aksoy and Yazlik [26].

One of the student errors is related to addition and subtraction with fractional especially if two fractions have a different number of the denominator. Based on the finding that integrated by [26], the common mistake that students make is writing the value of denominator of the fraction with the largest denominator value to the result statement and directly adding up the value of denominator of a fraction without equalizing the denominators. Figure 2 shows the mistake that student makes during the addition with the fraction. Next, students were asked to answer a question related equivalent fraction using model given by [26] and the success rate has increased compared to the previous question.

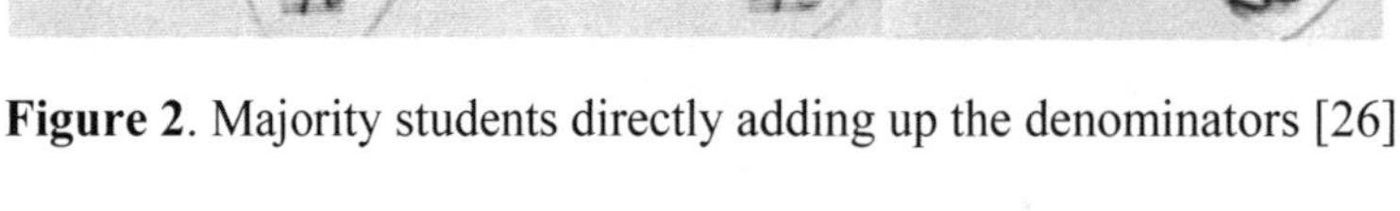

Figure 2. Majority students directly adding up the denominators [26].

2.1.3. Game

Game is an activity or a sport that one engages in for pleasure or amusement and the activity is in the form of competitive activity or sport that played according to the rules [5]. The game also can be defined as an activity that has rules, has a victory condition and it requires at least one player [6]. The digital game is a game that played using electronic devices such as computers, videogame console, and handheld device [7]. The game played on handheld devices such as smartphones, tablets or personal digital assistants (PDA) is known as handheld gaming or mobile game. A mobile game is defined as a game that embedded, downloaded or networked in handheld devices such as mobile phones, tablet and portable consoles [8]. In recent years, the number of games for mobile devices has exploded and can be easily downloaded and installed from the App Store for iOS devices and Play Store for Android devices.

2.1.4. Game Genre

There are various types of game genres that have been splintered off over the years. A game genre is used to classify the style of gameplay. Table 1 shows the game genre and its description [6], [9].

Table 1. Game genres and description.

Game Genres	Description
Action	A game that requires hand eye coordination to play. This genre has several subgenres including: • Action-adventure: this subgenre features an emphasis on item collection, puzzle solving and long-term story related goal. • Fighting: a game consists of two or more opponents battle in arena settings. Fighting games are distinguished from action games for the depth of the player controls. • Platformer: a game often features a mascot character jumping, shooting, fighting through challenging "platform" environment. • Stealth: a game with emphasis on avoiding enemies instead of fight against the enemies.
Adventure	A player is the protagonist of a story in the adventure game and the player needs to do the puzzle solving, item collection and inventory management. Adventure games often have a storyline with significant dialog. This genre has several subgenres including: • Role-playing game (RPG): this subgenre allows the player to choose a character class and increase the character's statistical abilities through treasure finding, combat and exploration. • Graphical adventure: a player uses a mouse or cursor to click to uncover clues and navigate around. • Massively multiplayer online role-playing game (MMORPG): an RPG game that can support hundreds of the players in one environment. • Survival/horror: a player often giving limited resources and attempt to survive a horror scenario.
Simulation	The game aims to simulate physical activities in real world. Some simulation game aim to provide simulations of forms of management such as city management and hotel management. Currently, the popular simulation game is The Sims series that allows the players to control the lives of the game characters. Sports game is one of the subgenres of simulation. • Sports games: a game stimulate the sporting experience such as football, snooker, fishing, tennis, golf etc.
Shooter	Shooter mainly focuses on firing projectiles at enemies. This genre has evolved to several subgenres that are distinguished by the camera view: • First person shooter: a shooter as seen from the player's perspective. • Shoot them up: arcade style shooters • Third person shooter: the camera is placed further behind the player
Racing	Racing games typically involving competing in a race against other players or time. Two subgenres can be identified: simulation and arcade.
Puzzle	A game requires the player to solve problems or puzzles, and can involve the exercise of logic, pattern matching, reaction time, memory, etc. The popular puzzle game is Tetris.
Education	A game designed for teaching purpose.
Traditional	A game represents computerized versions of board, card games, word, chess, mah-jongg, checkers etc.

2.2. Quality Requirements for Educational Game

According to [10] and [11], there are various quality requirements for the educational game that need to be considered, as follows:

- Learning Goal. The educational game should clearly define the learning goals of the game without skipping the fun elements. Besides, the game should give opportunities for the players to practice.
- Contains a full spectrum of learning principles. The game should provide direct feedback to the user. For instance, the audio of the gameplay important role to give feedback and provide instructional support. Besides, the game should create real-world problems that are relevant to the players.
- Evoke positive emotions. The game should have an attractive game design, interesting storyline or gameplay to guarantee that learners have fun. Furthermore, the game should provide positive feedback for learner's accomplishments so that the learners feel satisfaction.
- Evoke and keep motivate. Besides the learning goal, the game goals of each stage of the game also need to be clearly defined in order to create a sense of mission and motivate the user to play the game until achieving the mission. The difficulty game level should adapt to leaner' abilities and skills.

2.3. The Advantages of Handheld Augmented Reality

A handheld is defined as a piece of electronic device that designed to be held easily in one or two hands [5]. As the definition of handheld mentioned, handheld AR is describing as an AR system that can be held by the user's hand. In recent year, AR technology started implemented in handheld devices such as the smartphone and tablet. The obvious advantage of handheld AR is it allows people to carry it to any places. In addition, it can also be operated while people are walking [12].

Besides, one of the key advantages is the handheld AR inexpensive compared to the Head-Mounted Display (HMD) devices. Many people already own handheld devices such as smartphones and tablets. Moreover, to launch the AR application only requires the three necessary hardware components such as a computer processor, a display and a camera [13]. These hardware components can be found in the majority of mobile phones sold today.

2.4. The Benefits of Augmented Reality in Education

In recent years, many researchers found that AR applications are able to improve learning performance and motivate students [14]. One of the benefits is through AR technology, students can master the specific subject content easier and faster, because students can explore the objects from various perspectives [16]. With AR, there are possibilities for learners to control the objects by manipulating the size, position, shape, color and other properties of the virtual objects [17]. This allows learners to get more understanding of the properties and behavior of objects and at the same time can improve the learner's creativity.

Besides, many studies in literature found that AR learning can be effective in raising the learning achievement level [15], [17]-[19]. Furthermore, AR application able to trigger student's curiosity, interest, and motivation in learning as AR can help achieve a sense of reality [18]. Another study provided by Sirakaya and Cakmak [15] stated that students had fewer misconceptions in AR learning. There are a lot of similar studies in the literature support that AR has the ability to eliminate the misconception of the student [17], [18], [20], [21]. Another advantage obtained from AR learning is students are encouraged to study independently and follow students' own pace with AR [16]. Students are more self-responsible in academic progress and able to learn individually and explore or accept new knowledge [22].

Last but not least, AR in learning able to reduce costs compared to traditional learning. According to Chen and Tsai [23] the cost like execute manpower and the cost of signing or renewing the courses can be eliminated with the AR learning. In short, AR is eligible to be used in the educational area and a lot of studied had proof the AR application able to improve the learning achievements, student's motivation and allow student-centered learning.

3. Methodology
This section explains the methodology used to guide the development of proposed game. Waterfall approach has been selected to construct the project. The flow of the project methodology is divided into four sequential phases.

3.1. Phase 1: Requirements Analysis for Educational Game
In the first phase, research has been conducted to analyse the requirements of educational game. Besides, some study also conducted to get the related data and information about the proposed project. For example, the issues of the traditional learning method and the issues of learning the topic of fractions, this information helped to define the current problem in society.

3.2. Phase 2: Designing the Mathematics Game Structure
The previous phase is used as a reference for this design phase. The game design of the proposed game including the gameplay, game storyline, game characters as well as game is described in this phase to ensure the proposed game is fun to play.

3.2.1. Game Design
- Game Name: Ali's Math Adventure
- Game Concept: Ali's Math Adventure is an adventure game with problem solving and education elements where the player needs to answer the mathematics questions involving fractions to unlock the collections and new levels.
- Game Genre: Ali's Math Adventure is under the adventure and education genre.
- Game Play and Mechanics: The player allows selecting the choice given by using the 2D touch screen interaction to click the choice on the handheld screen. Figure 3 shows the game Mechanics.
- Game Story: There are two chapters in this proposed game. Both chapters have its own story to describe Ali's adventure journey.
- Game Level: The game difficulty follows the syllabus listed in Kurikulum Standard Sekolah Rendah (KSSR) Mathematics [24]. KSSR is the education system introduced by the Ministry of Education in 2011.

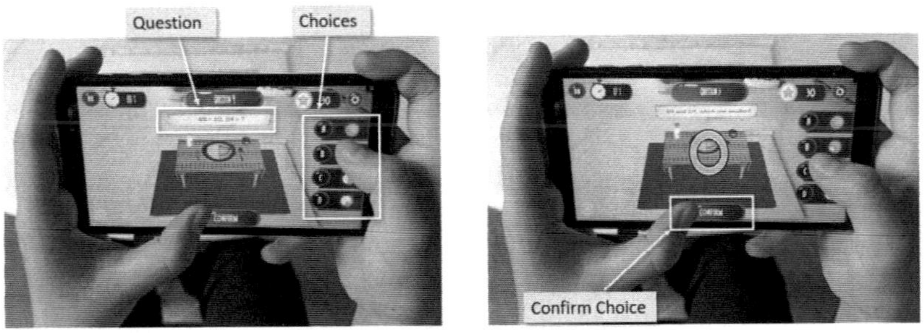

Figure 3. Game Mechanics.

3.3. Phase 3: Game Development and AR Implementation
In the third phase, Unity3D engine is used to develop the proposed game, and the target platform is Android handheld with the Android version 4.4 or above. After the game development, Vuforia Software Development Kit (SDK) integrated into the game to allow the interaction between the real-world environment with the three-dimensional virtual objects. Initially, the designed marker or image target uploaded to Vuforia Target Manager for processing and adding the image target to the database.

Figure 4 shows the image target used for this project. The implement process will be properly discussed in next section.

Figure 4. Image marker for the project.

3.4. Phase 4: Game Testing and Evaluation

In the game testing phase, functionality testing including the user acceptance testing method also known as the black box testing method was used to examine and determine the game developed to match the game requirements and specifications. The actions of the user were recorded for further evaluation and analysis. In order to evaluate the usability of the game, the suggestions or feedback given by the testers collected through pre-experiment and post-experiment questionnaire.

4. Implementation

The development of the mathematics game on handheld devices is carried out based on the game design stated. The AR tracking is then implemented to the game using the Vuforia AR Software Development Kit (SDK).

4.1. Develop AR Mathematics Game on Handheld Devices

Unity3D game engine is the main tool in the development of the proposed game and the targeting platform is the Android handheld devices. Initially, the Three - Dimensional (3D) objects are obtained from the online free open source model store such as Unity Asset Store. Those models are imported into the game scene and its position and scale been modified to a suitable position and scale.

A *GameController* class is scripted to handle the game logic. For instance, *GameController* randomly displays the question, receives the player's input, checks if the time limit reach, checks if the player's selection is correct, gives instant feedback to the player and calculates the score gained.

4.2. Implementation of Augmented Reality Tracking

The AR tracking method used in this project is a feature-based tracking technique, also known as an image-based tracking technique. This technique uses a camera to detect and track the features that are found in the two-dimensional (2D) image itself. To integrate the AR tracking in the project, Vuforia AR SDK is imported to the Unity3D game engine.

Figure 5 shows the workflow of the feature-based tracking using Android handheld camera with Vuforia SDK. The initial step is the Android handheld's camera capture the image marker and Vuforia SDK search for the input markers. Once the image marker identified, the Vuforia SDK match the tracked image or feature points with the Vuforia database. Once the matching is found, the Vuforia starts to calculate the viewpoint of the virtual object including the position and orientation of the virtual object on the real environment. Lastly, the three-dimensional virtual object displays on the Android handheld screen. The Vuforia SDK continues to track and display the virtual object on the screen until the feature marker is away from the camera's field of view.

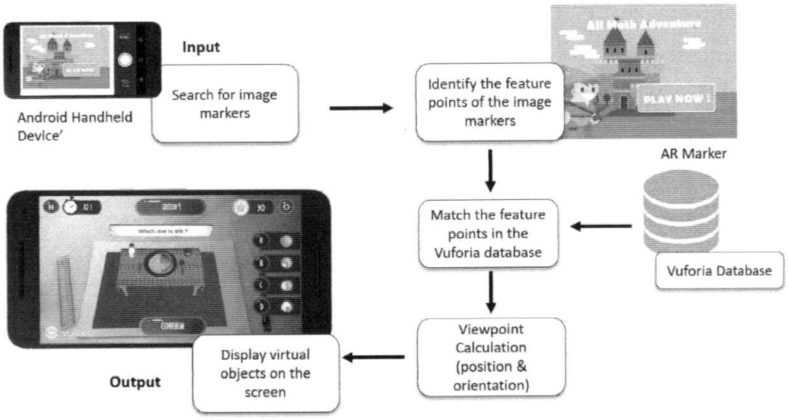

Figure 5. Feature-based tracking workflow using Android handheld camera with Vuforia SDK.

5. Results

The first objective of this project is related to the requirements analysis of related educational games. This was fulfilled in literature review as the requirements were studied based on past research papers. This serves as a guide to deliver a good quality educational game. Next, the second objective is related to development of the Mathematics game for primary school children using handheld device. Waterfall approach was used to manage the development progress by separating the task by phases and also maintained the development pace based on the proposed timeline. The third objective is related to implementation of Augmented Reality Technology into the Mathematics game. This objective has been fulfilled in the implementation stage of the project, including the implementation of AR tracking in the handheld device using Vuforia Software Development Kit (SDK).

The result of this project is a prototype of a learning mathematics game in an Android handheld device with the implementation of Vuforia AR Software Development Kit (SDK). The prototype is an educational game for primary school children and the subject Mathematics has been chosen for learning purpose. This prototype features Mathematics multiple choices questions with Two-Dimensional (2D) touch screen input to display the Three-Dimensional (3D) models on the AR marker in the Android handheld devices. Figure 6 shows the result of AR Mathematics game in handheld device.

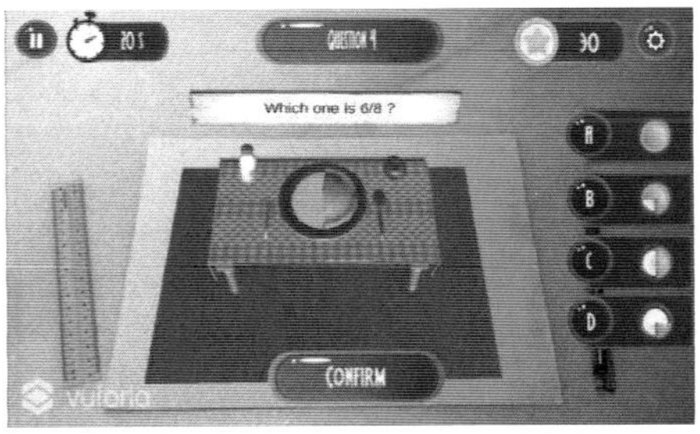

Figure 6. AR Mathematics Game.

6. Testing and Evaluations

6.1. Experimental Setup

The experiment has been set up as shown in Figure 7. This experiment requires a single-player, an android handheld device installed with the prototype's Android Package (APK) and also a printed A4 feature-based marker. The marker is placed on a flat surface, while the player requested to be seated in front of the marker. The handheld device is held by the player and the player can interact with the game prototype on a handheld device screen. The handheld device's camera used to track the image marker and render the game models.

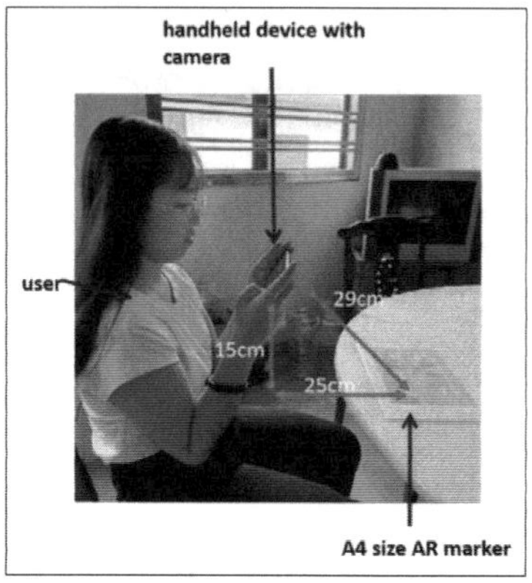

Figure 7. Experiment Setup.

6.2. Usability Testing

3 target users (3 female) and 10 non-target users (5 male and 5 female) were involved in this experiment and the selected number of participants is related to what was done in previous studies. The number of primary school children involved in the testing is limited due to the COVID- 19 outbreaks. Therefore, the testing is conducted by small sample of primary school children as the target users and several non-target users. The information and procedure of the testing has been provided to the parent or guardian of the target testers in order to get the permission for the testing. Testers could ask questions if they face any issues during the testing progress. For non-target users, any volunteers within the age from 18 to 24 years old as this aged is the largest group of handheld mobile users based on Nielsen [25]. Each participant is given around fifteen minutes to explore the prototype individually by following the experiment setup as discussed. The pre-experiment and post-experiment questionnaires were distributed to collect the participants' background, feedback and suggestions.

6.2.1. Evaluation Result for Non-Target Users

Based on the result shown in pre-experiment questionnaire for non-target users, 80% of respondents do experience with Augmented Reality (AR) and Handheld Augmented Reality (HAR). However, only 40% of the respondents had experience in learning Mathematics game on handheld devices. Based on the results shows in the post-experiment questionnaire for non-target users, most of the respondents gave positive feedback and also feel satisfied with the prototype. All respondents responded that the game user interface is attractive and the game is fun to play. Besides, one respondent gave a good reflection on the attractive game storyline and the game flow. Moreover, most of the respondents responded that AR tracking is effective and accurate and also the game can be play smoothly without facing any issue or evolved the feeling of frustrated. About the time given in each game level, only 10% of respondent feel it is not sufficient and suggested to increase the number of time while the rest 90% of respondents responded that the time given is sufficient for them.

6.2.2. Evaluation Result for Target Users

The result for target users in pre-experiment questionnaire shows that none of the respondent experience with AR and HAR. Only one respondent experience on learning Mathematics game using handheld device. Based on the results showed in the post-experiment questionnaire for target users, all respondents satisfied with the game prototype and gave positive feedback for the game user interface. Moreover, most of the respondents responded that AR tracking is effective and accurate and also the game can be play smoothly without facing any errors. About the time given in each game level, two out of three respondents suggested increasing the number of times. All respondents also agreed that the feedback of the game is clear after completed answered the question. In short, these results showed that the game prototype able to engage the children to play the game and practice the Mathematics questions independently.

7. Conclusion

Overall, this project has successfully produced a learning Mathematics game using HAR for primary school children. The proposed game is specifically designed for primary school children to strengthen the knowledge on the topic fractions and as an alternative method in learning Mathematics. Based on the testing results shown in both usability testing and black box testing, the game prototype with AR features is able to engage the children to play the game and trigger their curiosity in learning process. Besides, children are also able to practice the Mathematics questions independently in their preferred pace. To further enhance this project in the future, this proposed Mathematics game can be enhanced by implementing the multiplayer features. Secondly, the number of questions for each game level can be increased and store the list of questions in the cloud database. Currently, the AR tracking in this Mathematics game relies on image marker, the virtual objects will only appear superimposed in the real environment if the image marker is identified. Instead of using feature-based tracking, AR tracking can be enhanced by using motion tracking in the future.

8. Acknowledgement

Special thanks to Universiti Teknologi Malaysia, School of Computing and all partners involved in supporting this project and making this project a success.

9. References

[1] Etuk N 2008 Educational Gaming-From Edutainment to Bona Fide 21st-Century Teaching Tool In: *Multimedia & Internet@Schools*
[2] Boyle E A, Macarthur E W, Connolly T M, Hainey T, Manea M, Kärki A and Van Rosmalen P 2014 A narrative literature review of games, animations and simulations to teach research methods and statistics **74** 1-14
[3] Ratnottar S 2019 Augmented Reality (AR) Trends: The Past, Present & Future Predictions For 2019.
[4] Lin C-Y, Chai H-C, Wang J-y, Chen C-J, Liu Y-H, Chen C-W, Lin C-W and Huang Y-M 2016 Augmented reality in educational activities for children with disabilities *Displays* **42** 51-4
[5] Stevenson A 2010 In: *Oxford Dictionary of English (3 ed.)* Oxford University Press)
[6] Ismail A W 2019 Game History Genre, week 5, SCSV4213 Lectures Notes. p PDF slides
[7] Ilomäki L and Kankaanranta M 2009 *Handbook of research on new media literacy at the K-12 level: Issues and challenges*: IGI Global) pp 101-18
[8] Jeong E J and Kim D J 2009 Mobile computing: Concepts, methodologies, tools, and applications: IGI Global) pp 289-95
[9] Hanna P 2019 Video Game Technologies, course CSC207, Java Games Programming.
[10] Hense J and Mandl H 2012 pp 19-26
[11] Trisnadoli A, Muslim I and Novayani W 2016 Software Quality Requirement Analysis on Educational Mobile Game with Tourism Theme *JSW* **11** 1250-7
[12] Lindegger R and Student E 2013 Handheld Augmented Reality. In: *Distributed Systems Seminar*
[13] Billinghurst M and Henrysson A 2006 Research directions in handheld AR *IJVR* **5** 51-8
[14] Bacca J, Baldiris S, Fabregat R and Graf S 2014 Augmented reality trends in education: a systematic review of research and applications
[15] Sirakaya M and Cakmak E K 2018 The Effect of Augmented Reality Use on Achievement, Misconception and Course Engagement *Contemporary Educational Technology* **9** 297-314
[16] Kiryakova G, Angelova N and Yordanova L 2018 The potential of augmented reality to transform education into smart education *TEM Journal* **7** 556
[17] Shelton B E and Hedley N R 2002 Using augmented reality for teaching Earth-Sun relationships to undergraduate geography students. IEEE)

[18] Fleck S and Simon G 2013 An augmented reality environment for astronomy learning in elementary grades: an exploratory study. In: *Proceedings of the 25th Conference on l'Interaction Homme-Machine*: ACM) p 14
[19] Fotaris P, Pellas N, Kazanidis I and Smith P 2017 A systematic review of Augmented Reality game-based applications in primary education. In: *Proceedings of the 11th European Conference on Game-Based Learning (ECGBL 2017),* p 181
[20] Rosenbaum E, Klopfer E and Perry J 2007 On Location Learning: Authentic Applied Science with Networked Augmented Realities *Journal of Science Education and Technology* **16** 31-45
[21] Yen J-C, Tsai C-H and Wu M 2013 *Augmented Reality in the Higher Education: Students' Science Concept Learning and Academic Achievement in Astronomy* vol 103
[22] Diegmann P, Schmidt-Kraepelin M, van den Eynden S and Basten D 2015 Benefits of Augmented Reality in Educational Environments – A Systematic Literature Review
[23] Chen C-M and Tsai Y-N 2012 Interactive augmented reality system for enhancing library instruction in elementary schools *Computers & Education* **59** 638-52
[24] Kementerian Pendidikan Malaysia 2015 Kurikulum Standard Sekolah Rendah (KSSR), Matematik, Dokumen Standard Kurikulum dan Pentaksiran
[25] Nielsen 2016 Millennials are Top Smartphone Users
[26] Aksoy N C and Yazlik D O 2017 Student Errors in Fractions and Possible Causes of These Errors *Journal of Education and Training Studies* **5** 219-33

IOP Conf. Series: Materials Science and Engineering **979** (2020) 012015 doi:10.1088/1757-899X/979/1/012015

Virtual reality learning environments for vocational education: a comparative study with conventional instructional media on two-stroke engine

F Sholichin[1*], N M Suaib[1,2*], D A Irawati[1], Sutiman[3*], M Solikin[3], A Yudantoko[3], A Yudianto[3], I W Adiyasa[3], A J Sihes[4], H A Sulaiman[5]

[1]School of Computing, Faculty of Engineering, Universiti Teknologi Malaysia, 81310 UTM Skudai, Johor, Malaysia.
[2]UTM Big Data Center, Ibnu Sina Institute of Scientific and Industrial Research, Universiti Teknologi Malaysia, 81310 UTM Skudai, Johor, Malaysia.
[3]Automotive Engineering of Education, Faculty of Engineering, Universitas Negeri Yogyakarta
[4]School of Education, Faculty of Social Science and Humanities, Universiti Teknologi Malaysia, 81310 UTM Skudai, Johor, Malaysia.
[5]Universiti Teknikal Malaysia Melaka, 76100 Durian Tunggal, Melaka

[1*]sholichin.fauzi@graduate.utm.my, [2*] haida@utm.my, [3*]sutiman@uny.ac.id

Abstract. Virtual reality technology has been adopted as a modern learning media in vocational education. Nowadays, the practical learning needs complete equipment and tools to achieve the goal study. However, the conventional media requirement is a challenge for academician for learning process, specifically in automotive and machine areas. Virtual reality provides a solution for this problem, but still needs proper analysis on the impact of usability between conventional and VR media to which each learning objective is achieved. The case study used in this research is a two-stroke engine, which is currently a rare item in the real world. We replicated this two-stroke engine in the developed VR system using Unity game engine and Oculus quest. A comparison on the usability effectiveness on conventional and virtual reality methods was conducted using cognitive walkthrough, which follows the four steps of learning objective training. The result shows that three out of four learning objectives are similar with conventional objective training. One of the process on preparation for disassembly machine has slight difference with conventional training; faulty engine due from environment elements (liquid, rusty iron) was not included.

1. Introduction

Virtual reality provides a technology, which increases the efficiency and quality of training in the Industry Revolution 4.0 and learning environments imitating a real-life process or situation [1]–[6]. The development of engineering and technology gives the new opportunities for improving the educational of methods process. Specifically, an engineering education is related with machine such as automotive, manufacturing, and industry. Therefore, there is a need for educational virtual technologies development, especially, in the technology classified as rare as the two-stroke engine.

This paper describes the study on a VR training of a two-stroke engine simulator as an alternative learning environment for vocational training. Findings from this study is hoped to shed some ideas for future research and design for suitable VR training system that can be used in the automotive area. The main findings of this study can be highlighted as follows; firstly, it is evident that a suitable VR system can be used as an alternative to learn the subject of two-stroke engine. Secondly, this paper outlines the technical information on the design of the 3D model as well as the user interaction to support a realistic two-stroke engine simulation. Third is the evaluation of virtual reality for learning two-stroke engine with instructional media using cognitive walkthrough. The gap between VR and instructional media forms the basis for further product development, which the usability method is suitable for filling the gap.

The rest of this paper is organized as follows: a review on related research and methods is presented in Section 2. A detailed description on the related design and modelling process for this study is shown in Section 3. Section 4 outlines the settings of the experiments, while Section 5 provides a cognitive walkthrough evaluation result between the VR learning system and traditional methods. Section 6 presents the threat and validity of this research. And finally, the main conclusions, limitations and future works is presented in the last section.

2. Background and related work

In the context of skill training, virtual reality offers many advantages used in education field [5], [7], [8]. The virtual environment may allow students to learn practically, which enhance the learning and processing skill [7], [9], [10]. Modern engineering education supposes that the learning process needs to combine harmoniously of fundamental theoretical training with the skills of practical work on solving the real problems of the industry [9], [11]. Especially in automotive, the learning process needs specific tools and vehicle to archives each learning goals. Virtual reality technology is one of a variety of learning domains to improve immersive learning environment in vocational study.

Providing effective training for future workers can present significant challenges, requires access to real equipment or physical mock-ups [4], [12], [13]. Therefore, the usability side is crucial in the learning process following a lesson plan of subject. The usability method used to evaluate between VR and instructional media is a cognitive walkthrough. The cognitive walkthrough is a usability walkthrough technique that focuses primarily on the ease of learning of a product [14]. Which, the group of evaluators in this research have been collaborate with automotive engineering of education students. The cognitive walkthrough can be used in the during of the development process, which a process namely requirements, conceptual design, detailed design, implementation [14]. However, in this research the focus is on the comparison in implementation process with follow up the skill and experience user.

Table 1. Document and Material Required for the Cognitive Walkthrough.

Description	Required
User profile	Automotive engineering of education students
Task list	Four step based lesson plan : 1. Checking tools and requirement 2. Preparation for disassembly of two stroke 3. Disassembly procedures of two stroke engines 4. Reassembly procedures of two stroke engine
Action sequence of task	Include following information : Task, Users, and Interface.
Problem reporting form	Tasks based on lesson plan
Representation of the UI	Prototypes and working feature

The document and material required for the cognitive walkthrough in this research is shown as Table 1 namely, user profile, task list, action sequence of task, problem reporting form, and representation user interface. Learning method between instructional media and virtual reality needs to be measured to see the comparison of the efficacy of this media in learning process. Therefore, the analysis of gap between user reactions between the real world training and virtual work training will be needed to evaluate the VR system.

3. Modelling of VR two stroke engine

Modelling of the environment and engine components were done using SOLIDWORKS and Blender. Whereas, the components that control interactivity, object interaction, user behaviour, were implemented using Unity. The full development of the VR system involves the use of C# which is supported by Unity to implement the algorithms for navigation and behaviour [15]. The production phases of VR two-stroke engines were described on the modelling process.

3.1. *The modelling process*

Modelling is an essential part of virtual reality system. The modelling phase was completed on the early phase of development. The virtual model of the two-stroke engine and workshop were created using SOLIDWORKS. The VR environment consists of these components; engine, room of workshop, and auto mechanic tools. The 3D workshop building is important to bring student engagement in the room of practical lesson. The two-stroke engine model was created following each component, which is shown on Figure 2.

Development phase was divided on two types of works. The first phase is the design work, which covers VR simulation concept and model content creation. The second is the work inside Unity environment on VR simulation development. The model content creation embraces 3D two stroke engine diagram and auto workshop diagram. Figure 1 gives detail of simulation production pipeline of development phase. The pipeline shows the task that needs to be completed before continuing the next step.

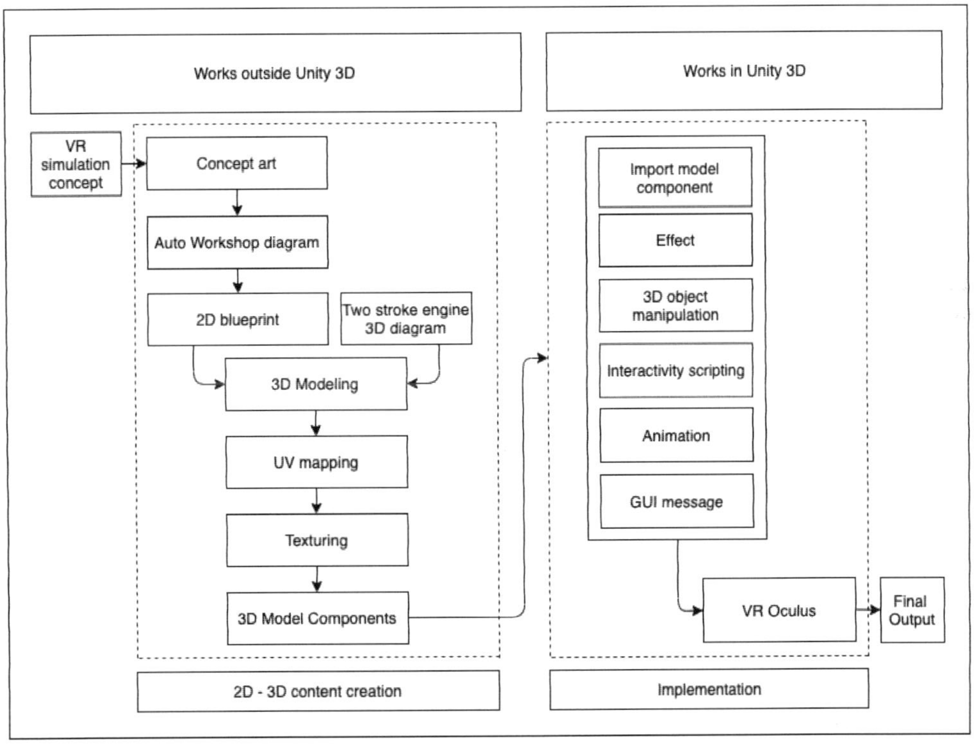

Figure 1. Simulator production pipeline for VR two stroke engine.

Figure 2 gives a diagram of the component of the two-stroke engine. The result of modelling shows as Figure 3 for engine, and Figure 4 for workshop.

IOP Conf. Series: Materials Science and Engineering **979** (2020) 012015 doi:10.1088/1757-899X/979/1/012015

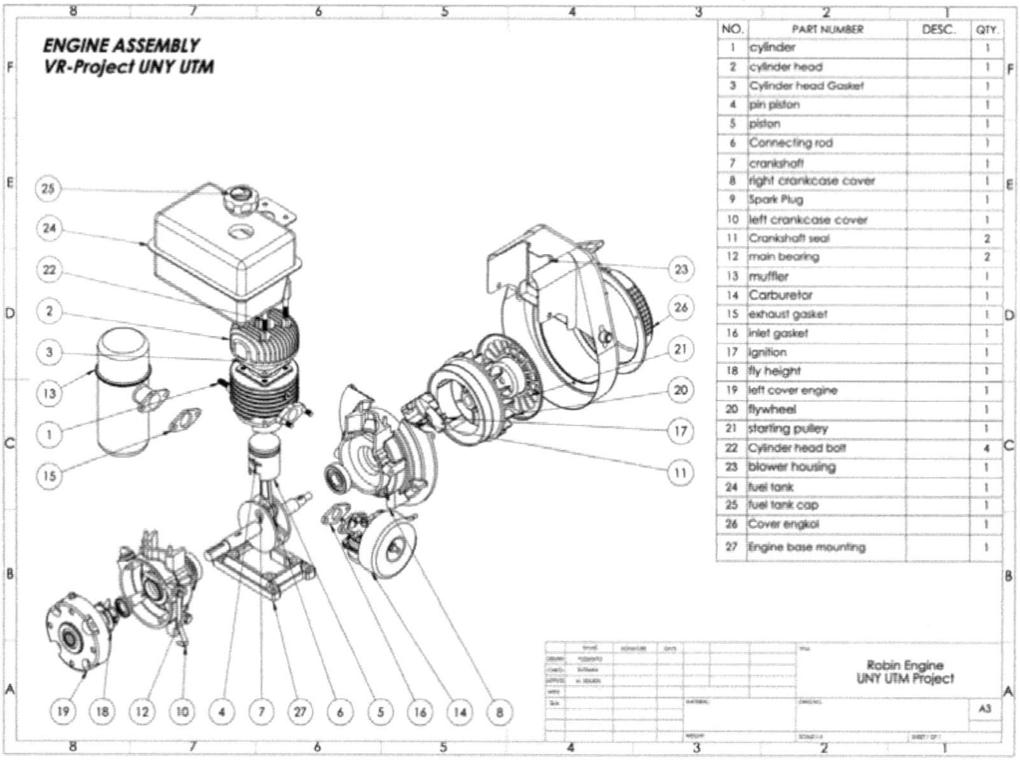

Figure 2. Modelling components.

Figure 3. Engine Assembly.

4. Experiment

4.1. Implementation

The developed VR learning system was tested on a notebook with 16GB of RAM, NVIDIA GTX 1080 graphic card and 2.80GHz Intel i7-7700 3.60 GHz (8CPUs) processor. The user experienced the two-stroke engine simulation via the Oculus Quest. The development of the VR system used Unity version 2019.x3 and SOLIDWORKS version 2020. Figure 4 and 5 shows VR simulation with oculus in the process of implementation.

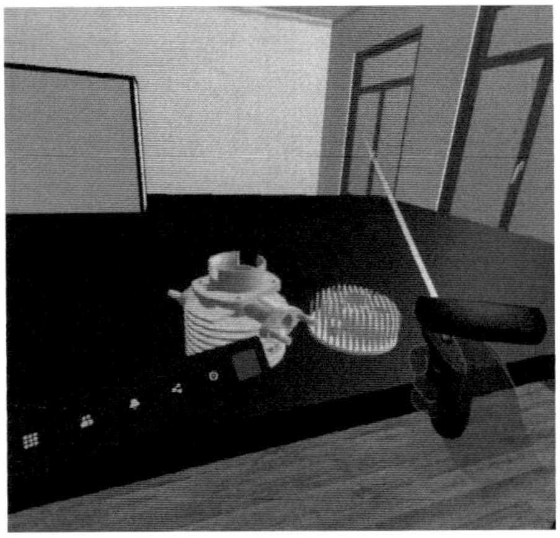

Figure 4. Simulation with controller.

Figure 5. Simulation with hand tracking.

4.2. Participants

Next, we will explain the steps in verifying the validity of the VR platform for the two-stroke engines simulation. A total of 20 participants were involved in the experiment; of whom these volunteers have auto mechanic skill and had experienced the conventional training beforehand. Participants were recruited among the students of automotive engineering course at Universitas Negeri Yogyakarta from different semesters. The students were selected based on the subject program, who graduated of automotive engineering education targeted to be teachers and academicians specializing in automotive areas. The characteristic of study participants is presented in Table 2.

Table 2. Sample characteristic.

Data	Values
Number of subjects	20
Male subject	20
Female subject	0
Minimum age	19
Maximum age	24
Mean age	22
Virtual reality game expertise	2 Yes (10 %) / 18 No (90%)

Furthermore, all of subjects are male, with less/no virtual game expertise (90%). 2 of subjects declared some experience using VR with their smartphone. Since all participants has the skill needed in this experiment, it is convenient to compare conventional instructional media and virtual reality on training of two-stroke engine.

5. Cognitive walkthrough evaluation

5.1. *Experiment description*

The participants were tested with an instructional media and virtual reality. The test is composed of four steps, namely (1) checking tools and equipment, (2) preparation for disassembly of two stroke, (3) Disassembly procedures of two stroke engines, and (4) Reassembly procedures of two stroke engine. Immediately in after completing each step, participants were asked to complete tasks training. The flowchart of training procedure shows at Figure 6.

IOP Conf. Series: Materials Science and Engineering **979** (2020) 012015 doi:10.1088/1757-899X/979/1/012015

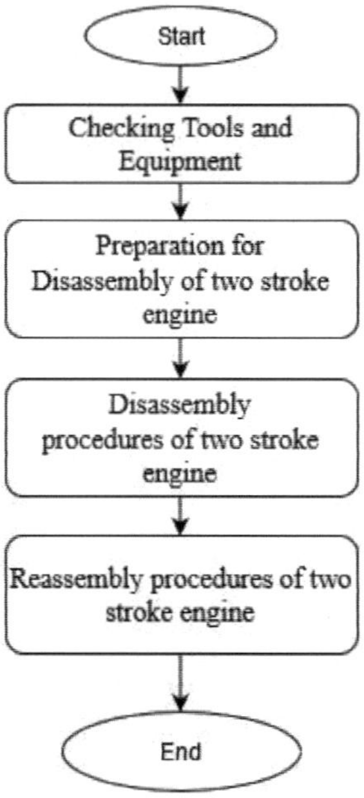

Figure 6. Flowchart of the training procedure.

5.2. Evaluation result

Usability evaluation is both an essential and challenging task in this research. Each of training parts have specific requirement based on the training module / curriculum. In this case, learning two-stroke engine involves four steps that were previously described on section 5.1. The first step is related to tools and equipment needed for student learning process. Table 3 gives a comparison result in the first step. Result shown between conventional method and virtual reality methods, each of tools and equipment fulfil 100%, which shows the similarity between both methods.

Table 3. Result comparison of Tools and equipment.

Tools and Equipment	Conventional Methods	Virtual Reality Methods
Wrench Set	YES	YES
Screwdriver Set	YES	YES
Torque Wrench	YES	YES
Special Tool; Flywheel Puller	YES	YES

Next step is a preparation for disassembly of two-strokes engine. Table 4 gives a detail comparison of conventional and virtual reality methods. The result shows a preparation for disassembly have a part that shows dissimilarity where the VR system cannot fulfill. The reason is for detailed material such as density of materials to classify a fragile object, any chemical elements that are not part of a main engine component, and other liquid machine that were not provided for the VR system. However, in this part, the step is more concise on VR system, in which a participant can complete faster and safer.

Table 4. Result comparison of Preparation for Disassembly.

Preparation for Disassembly	Conventional Methods	Virtual Reality Methods
1. When disassembling the engine, memorize where and how each part is assembled in order to be able to reassemble it correctly.	Yes	Yes
2. Take care not to damage packings and gaskets, which are fragile.	Yes	NO (Object not fragile)
3. In order to prevent missing and misplacing, group related parts together parts together, tentatively assembling them, immediately after disassembled each sub-assembly.	Yes	No (Handle by feature on VR)
4. Handle the disassembled part carefully, and wash them in kerosene.	Yes	No (Not required)
5. Use the correct tools in the correct way.	Yes	Yes
6. Standard tools required for disassembly and reassembly. a. Work table b. Washing pan c. Disassembling tools d. Washing oil e. Emery paper, cloth	Yes	Yes
7. Before starting to disassembly, the engine, drain fuel and lubricant.	Yes	No (Machine and Engine always ready)

Table 5. Result comparison of Disassembly Procedures.

Disassembly Procedures	Conventional Methods	Virtual Reality Methods
1. Fuel tank and fuel tank bracket a. Disconnect fuel pipe between strainer and carburetor at carburetor side b. Remove mounting bolts and detach fuel tank from cylinder head or cylinder	YES	YES
2. Air cleaner a. Remove air cleaner cover and element b. Unscrew two bolts which clamped air cleaner base plate to carburetor and remove base plate	YES	YES
3. Carburetor a. Remove governor rod and rod spring from carburetor b. Remove carburetor from cylinder block	YES	YES
4. Governor lever a. Remove governor lever from governor shaft b. Remove governor spring from control lever c. Control lever and stop plate can be remove by losing wing-nut or bolt, but do not disassembly unless it necessary	YES	YES
5. Muffler A. Unscrew bolts or nut and remove it from cylinder	YES	YES
6. Blower housing a. Disconnect stop button wires from connector	YES	YES

b. Unscrew bolts and remove blower housing from crankcase		
7. Recoil starter a. Remove recoil starter from blower housing by unscrew the bolts b. Remove starting pulley from flywheel by losing bolts clamped	YES	YES
8. Starting pulley and magneto a. Remove starting pulley from flywheel b. Remove flywheel from crankshaft c. Remove spark plug cap from high tension cable of ignition coil and remove ignition coil from crankcase d. Take off the point cover and remove contact breaker and condenser, from crankcase	YES	YES
9. Cylinder head a. Remove spark plug from cylinder head b. Unscrew mounting special bolts and remove cylinder head from cylinder c. Remove cylinder head gasket from cylinder	YES	YES
10. Cylinder a. Remove cylinder from crankcase by removing nut and spring washer b. Remove cylinder gasket c. Remove intake pipe from cylinder	YES	YES
11. Piston a. Remove both sides piston pin clips b. Pull out piston pin and disconnect the needle bearing from small end of connecting rod c. Remove piston rings from piston by expanding the open ends of rings	YES	YES
12. Governing plate a. Pull out governor plate, governor sleeve, washer and clip from crankshaft b. Remove woodruff key	YES	YES
13. Governing plate A. Pull out governor plate, governor sleeve, washer and clip from crankshaft B. Remove woodruff key	YES	YES

Table 6. Result comparison of Reassembly Procedures.

Reassembly Procedures	Conventional Methods	Virtual Reality Methods
1. Fuel tank and fuel tank bracket a. Reassembly with reverse sequence based on disassembly procedures correctly	YES	YES
2. Air cleaner a. Wash elements before reassembly: outer and inner elements with gasoline, soak them into the mixed oil of 2-4 gasoline and 1 mobile oil b. Reassembly with reverse sequence based on disassembly procedures correctly	YES	No (Wash element part)
3. Carburetor A. Reassembly with reverse sequence based on disassembly procedures correctly	YES	YES

IOP Conf. Series: Materials Science and Engineering **979** (2020) 012015 doi:10.1088/1757-899X/979/1/012015

4. Governor lever a. Reassembly with reverse sequence based on disassembly procedures correctly b. Assembly shall be made securely including rotation adjustment	YES	YES
5. Muffler a. Reassembly with reverse sequence based on disassembly procedures correctly	YES	YES
6. Blower housing a. Reassembly with reverse sequence based on disassembly procedures correctly	YES	YES
7. Recoil starter a. Reassembly with reverse sequence based on disassembly procedures correctly	YES	Yes
8. Starting pulley and magneto a. Reassembly with reverse sequence based on disassembly procedures correctly b. Reassembly should include breaker point adjustment and timing adjustment	YES	YES
9. Cylinder head a. Clean carbon from combustion chamber and dirt from between the cooling fins of cylinder head. Check its mounting face for distortion b. Use new cylinder head gasket c. Cylinder head fin must be placed in parallel with crankshaft. Fasten with recommended fastening torque d. Apply the standard fastening torque for spark plug e. Reassembly with reverse sequence based on disassembly procedures correctly	YES	No (Clean carbon from combustion chamber)
10. Cylinder a. Clean carbon deposit from cylinder head and combustion chamber b. Replace cylinder gasket to the new one c. Intake cylinder should be positioned to the left against view from blower side. Apply oil to piston ring and cylinder walls, then after assembled the cylinder securely, make sure if the crankshaft rotates smoothly d. Apply the standard fastening torque of cylinder head and spark plug e. Reassembly with reverse sequence based on disassembly procedures correctly	YES	YES
11. Piston a. In reassembly of piston rings, if an expanding tool is not available, install rings by placing the open end of the ring on first land of piston, spread ring only far enough to slip over the piston and carry it into correct groove b. In reassembling the piston, position f mark of piston to blower side and reassemble the piston and connecting rod by firmly striking the piston pin and needle bearing c. Assemble piston pin clip	YES	YES

d. Be sure that piston and connecting rod moves smoothly after reassembled		
e. Reassembly with reverse sequence based on disassembly procedures correctly		
12. Governing plate	YES	YES
a. Reassembly with reverse sequence based on disassembly procedures correctly		

The third and fourth steps are disassembly and reassembly of two stroke engine, which is almost similar between conventional and virtual reality methods. The differences are only in the use of the chemical and liquid engine in conventional workshop, but not on the VR system. Therefore, this part of the step was skipped on VR system. Overall, the VR system can fulfil the training on two-stroke engine instructions, based on actual steps on conventional methods. The result shows a VR system for training, specifically in automotive and machine area not only focus on engines components, but also needs detail of a component of virtual environment, such as liquid, dusk, that any material on used machine. Therefore, improve the similarity of between conventional and virtual reality methods for simulator training goal.

6. Threats to Validity

This paper has some limitations acknowledged, which may threaten its legitimacy. The potential threat of this paper is associated with less case studies. The case study use in this research is on a two-stroke engine machine, which is chosen because of the difficulty to find this type of engine in good condition nowadays. However, the specific major such as automotive engineering of education curriculum still needs to be delivered to students.

7. Summary

In this study, we developed a two-stroke engine simulator for student learning and training tools with new VR technology. Our proposed VR simulation developed for the current study provides an alternative for participants to learn the two-stroke engine with a four-step learning process. The evaluation results mainly show that the step on VR system has similarity with the conventional learning process. Moreover, the proposed VR systems has the potential of reducing total cost involved for the training institution compared to the conventional method, and improved safety side from any fragile part and hazardous chemicals. The only downside to current VR system is the limitation on the step of preparation for disassembly engine. Since the VR system has not included details on the objects (eg. density, chemical, rust), the risk due to errors in the object cannot be simulated successfully. The use of a VR two-stroke engine simulator however has many benefits over conventional learning methods. Most important advantage is it offers a learning environment with increased safety level and minimum risk to the participants.

The authors involved in this study are currently studying the efficacy of VR technology adoption into learning/training on various automotive machines such as motorcycle and car engine operations. It is also interesting to identify the factors that provide gaps for details of the virtual environment such as fluids, materials, electricity in machines that do not include diagrams but occur in real-world training. By identifying the factors that have been shown to influence each phase in the step-by-step training process, future researchers can begin to develop conceptual problem-solving models in the VR learning process.

7.1. Limitations and future works

We identified some limitations of the present work that require attention in future research. The present VR system is able to simulate the two-stroke engine scenario mainly by visual stimulation. However, are also factors characterized by fluid and chemical materials as well as by the heat and airflow that is normally generated by the engine – these need some improvement to appear on the VR simulation. Future studies need to include natural stimulants such as olfactory and heat based on engine condition. It is hoped that this will increase the level of realism on user experience so that in turns will increase the effectiveness of the VR two-stroke engine simulation.

References

[1] I. Makarova, A. Boyko, K. Shubenkova, A. Pashkevich, and I. Giniyatullin, "Virtual laboratories: Engineers' training for automotive industry," *ICETA 2019 - 17th IEEE Int. Conf. Emerg. eLearning Technol. Appl. Proc.*, pp. 505–511, 2019, doi: 10.1109/ICETA48886.2019.9040074.

[2] D. Rath, I. Satpathy, and B. C. M. Patnaik, "Augmented reality (Ar) & virtual reality (Vr)-a channel for digital transformation in industrialization fostering innovation & entrepreneurship," *Int. J. Innov. Technol. Explor. Eng.*, vol. 8, no. 10, pp. 3228–3236, 2019, doi: 10.35940/ijitee.J1167.0881019.

[3] M. Carfagni *et al.*, "Fast and Low Cost Acquisition and Reconstruction System for Human Hand-wrist-arm Anatomy," *Procedia Manuf.*, vol. 11, no. April 2020, pp. 1600–1608, 2017, doi: 10.1016/j.promfg.2017.07.306.

[4] X. Wu and G. Fei, "Research of virtual reality technology in automotive engine assembly teaching," *Proc. - 2011 6th IEEE Jt. Int. Inf. Technol. Artif. Intell. Conf. ITAIC 2011*, vol. 1, pp. 167–169, 2011, doi: 10.1109/ITAIC.2011.6030177.

[5] J. A. di Lanzo, A. Valentine, F. Sohel, A. Y. T. Yapp, K. C. Muparadzi, and M. Abdelmalek, "A review of the uses of virtual reality in engineering education," *Comput. Appl. Eng. Educ.*, vol. 28, no. 3, pp. 748–763, 2020, doi: 10.1002/cae.22243.

[6] W. X. Quevedo *et al.*, "Virtual reality system for training in automotive mechanics," *Lect. Notes Comput. Sci. (including Subser. Lect. Notes Artif. Intell. Lect. Notes Bioinformatics)*, vol. 10324 LNCS, pp. 185–198, 2017, doi: 10.1007/978-3-319-60922-5_14.

[7] M. Akçayır and G. Akçayır, "Advantages and challenges associated with augmented reality for education: A systematic review of the literature," *Educ. Res. Rev.*, vol. 20, pp. 1–11, 2017, doi: 10.1016/j.edurev.2016.11.002.

[8] O. T. Laseinde, S. B. Adejuyigbe, K. Mpofu, and H. M. Campbell, "Educating tomorrows engineers: Reinforcing engineering concepts through Virtual Reality (VR) teaching aid," *IEEE Int. Conf. Ind. Eng. Eng. Manag.*, vol. 2016-Janua, no. December, pp. 1485–1489, 2016, doi: 10.1109/IEEM.2015.7385894.

[9] I. Makarova, R. Khabibullin, E. Belyaev, and A. Bogateeva, "The application of virtual reality technologies in engineering education for the automotive industry," in *2015 International Conference on Interactive Collaborative Learning (ICL)*, 2015, pp. 536–544, doi: 10.1109/ICL.2015.7318086.

[10] M. Belani, "Evaluating Virtual Reality as a Medium for Vocational Skill Training," pp. 1–8, 2020, doi: 10.1145/3334480.3375027.

[11] K. Tcha-Tokey, E. Loup-Escande, O. Christmann, and S. Richir, "A questionnaire to measure the user eXperience in immersive virtual environments," *ACM Int. Conf. Proceeding Ser.*, 2016, doi: 10.1145/2927929.2927955.

[12] D. W. Carruth, "Virtual reality for education and workforce training," *ICETA 2017 - 15th IEEE Int. Conf. Emerg. eLearning Technol. Appl. Proc.*, 2017, doi: 10.1109/ICETA.2017.8102472.

[13] C. Sagnier, E. Loup-Escande, D. Lourdeaux, I. Thouvenin, and G. Valléry, "User Acceptance of Virtual Reality: An Extended Technology Acceptance Model," *Int. J. Hum. Comput. Interact.*, vol. 36, no. 11, pp. 993–1007, 2020, doi: 10.1080/10447318.2019.1708612.

[14] C. Wilson, "Chapter 4 - Cognitive Walkthrough," in *User Interface Inspection Methods*, C. Wilson, Ed. Boston: Morgan Kaufmann, 2014, pp. 65–79.

[15] "Unity Documentation," 2020. .

IOP Conf. Series: Materials Science and Engineering **979** (2020) 012016 doi:10.1088/1757-899X/979/1/012016

Enhanced Canny edge detection for Covid-19 and pneumonia X-Ray images

S K T Hwa[1*], A Bade[1] and M H Ahmad Hijazi[2]

[1]Faculty of Science and Natural Resources, Universiti Malaysia Sabah, Malaysia
[2]Faculty of Computing and Informatics, Universiti Malaysia Sabah, Malaysia

[*]stefanuskieu@gmail.com

Abstract. In image processing, one of the most fundamental technique is edge detection. It is a process to detect edges from images by identifying discontinuities in brightness. In this research, we present an enhanced Canny edge detection technique. This method integrates local morphological contrast enhancement and Canny edge detection. Furthermore, the proposed edge detection technique was also applied for pneumonia and COVID-19 detection in digital x-ray images by utilising convolutional neural networks. Results show that this enhanced Canny edge detection technique is better than the traditional Canny technique. Also, we were able to produce classifiers that can classify edge x-ray images into COVID-19, normal, and pneumonia classes with high accuracy, sensitivity, and specificity.

1. Introduction

Edge detection is a fundamental image processing technique for locating the borders of objects in images. It works by identifying disjointedness in brightness. One of the edge detection algorithms is Canny.

Edge detection methods can produce a new set of features. Edges are noted by serious or local changes in the image, and edges normally occur on the borders between two distinct areas in an image. Edges are important features for analysing digital images [1], [2]. Numerous edge detection techniques are available.

Traditional gradient-based edge detection techniques like Robert, Prewitt, Sobel were among the first algorithms proposed, but they are sensitive to noise conditions and did not give sharp edges [3]. Some flaws of Laplacian operators' flaws are a high likelihood of identifying false edges and huge localization error at rounded edges [2]. Canny proposed a multi-stage edge detection algorithm, and this technique is deemed as the optimal edge detection technique for images that with noise [4]. Canny proposed the optimum edge detection procedure that decreases the likelihood of identifying fake edges. Canny edge detection also gives sharp edges. However, the traditional Canny edge detection also has flaws. One of which is that the even though Gaussian filtering removed the noise, the important edges were also removed or smoothed [5]. This weakened the edges and increases false edge detection.

X-ray is a type of medical image in which a part of the patient's body is exposed to a minor dose of ionizing radiation, producing pictures of the body's inside. X-ray helps physicians treat and diagnose medical conditions [6]. In this research, we present an enhanced Canny edge detection method. Furthermore, we also apply the proposed edge detection technique for COVID-19 and pneumonia detection in digital chest x-ray images by utilising convolutional neural network (CNN).

IOP Conf. Series: Materials Science and Engineering **979** (2020) 012016 doi:10.1088/1757-899X/979/1/012016

2. Literature Review

In this section, the concepts of morphological contrast enhancement and Canny edge detection were presented. The usage of deep learning techniques in COVID-19 and pneumonia detection was also discussed.

2.1. Morphological Contrast Enhancement

Morphology image processing are processes that manipulate images based on structuring elements. Morphological operations accept an image as input, apply a structuring element to the image, and outputs the resulting image. The pixel of the output image is determined from a comparison between the corresponding pixel of the input image with its neighbours.

In morphological contrast enhancement, a filter is applied. If the central pixel is closer to the local maximum, this filter replaces it with the local maximum. Else, the central pixel is replaced by the local minimum. Figure 1 shows the pseudocode implementing morphological contrast enhancement.

```
Get original pixel value, Po
Find local maximum, Pmax
Find local minimum, Pmin
Compute difference between original pixel value and local maximum, D1
where D1 = |Po - Pmax|
Compute difference between original pixel value and local minimum, D2
where D2 = |Po - Pmin|
If D1 < D2
        Po = Pmax
Else
        Po = Pmin
```

Figure 1. Pseudocode for morphological contrast enhancement.

2.1.1. Canny Edge Detection

The Canny edge detection technique is a multi-step procedure. A good edge detection algorithm should be able to detect edges with a low error rate, localize edge point on the centre of the edge, and mark an edge only once.

The Canny filter is a multi-stage edge detector. First, noise reduction was performed by implementing Gaussian smoothing. Then, the intensity gradient of the image is computed so that non-maximum suppression can be employed. Likely edges are reduced to 1-pixel curves by discarding non-maximum pixels of the gradient magnitude. Lastly, hysteresis thresholding is performed to determine whether or not to keep the edge pixels. The detailed explanations for Canny edge detection are as follows:

Step 1: Noise Reduction

Before attempting to locate and identify edges, noise reduction is carried out to remove noise in the original image. The noise here refers to a random variation of color or brightness in the image. Canny edge detection uses the Gaussian filter to perform noise reduction. A Gaussian mask is calculated and applied using standard convolution methods. In short, a Gaussian filter was used to smooth the image to reduce unwanted textures and details.

Step 2: Calculating Intensity Gradient

After noise reduction, the image gradient is calculated to obtain the edge strength. The edge strength is also known as the absolute gradient magnitude. A 2D spatial gradient calculation is performed using the Sobel operator. This operator uses two convolutional masks to estimate the x-direction gradient and y-direction gradient. We can find the edge gradient as follows:

$$G = \sqrt{G_x^2 + G_y^2} \tag{1}$$

Once the x and y gradients are known, the following formula can be used to find the edge direction:

$$\theta = \tan^{-1}\left(\frac{G_y}{G_x}\right) \tag{2}$$

Step 3: Non-maximum Suppression

After obtaining the gradient direction as well as the gradient magnitude, non-maximum suppression was performed. This was done to remove pixels that might not be a part of an edge. Each pixel is inspected to see if it is a local maximum in its neighbourhood in the direction of the gradient. This will result in a binary image with "thin edges".

Step 4: Hysteresis Thresholding

This step classifies edges into real edges and fake edges. Two threshold values are needed, which are min-value and max-value. All edges with an intensity gradient greater than the max-value are considered as "sure-edges", thus kept, and any edges under the min-value are considered as fake edges, thus removed. The remaining lines that are between these two threshold values are categorised into real edges and fake edges based on their connection. If they are joined to "sure-edge" pixels, they are counted to be part of an edge. If not, they are also removed. Figure 2 provides some examples.

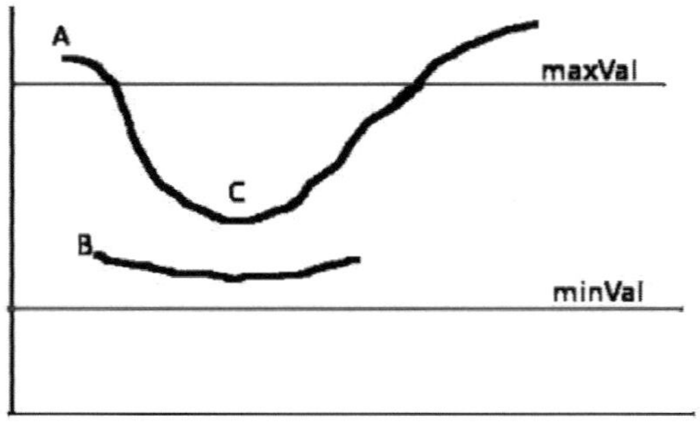

Figure 2. Hysteresis Thresholding.

Edge A is over the max-value, so it was counted as "sure-edge". Even though edge C is under max-value, it is joined to edge A, therefore it was counted as a proper edge and the full curve was obtained. Even though edge B is over the min-value and also in the same area as edge C, it is not joined to any "sure-edge", so is was removed. Therefore, it is highly imperative to choose max-value and min-value appropriately to obtain better result.

This step also assume that edges are long lines. Therefore, small pixels noises were discarded. Finally, an image with strong edges is produced.

2.1.2. CNN for COVID-19 and Pneumonia Detection

COVID-19 is diagnosed by the reverse transcription polymerase chain reaction (RT-PCR) test [7]. COVID-19 is caused by severe acute respiratory syndrome coronavirus 2 (SARS-CoV-2), and RT-PCR can detect SARS-CoV-2 strain. To perform the RT-PCR test, a swab is inserted into the nostril of a patient and gently moved into the nasopharynx to collect nasal secretions sample. The sample obtained will be used for the RT-PCR test. However, radiologists can also use chest x-ray images to identify COVID-19 infection [8]. On the other hand, pneumonia can also be detected from a chest x-ray [9]. However, COVID-19 share similar imaging features with pneumonia [10]. Therefore, it is imperative to speedily and accurately differentiate people infected with COVID-19 and people infected with pneumonia.

Deep learning methods were proposed to carry out classification between healthy lungs, COVID-19 infected lungs, and pneumonia infected lungs [11]–[15]. [11] used CNN to classify computed tomography scan images into normal COVID-19 and Influenza-A-viral-pneumonia, infected lungs. [12] trained a CNN and used data augmentation such as horizontal flip, rotation, translation, and intensity shift to perform 3-class classification (normal, pneumonia and COVID-19) on chest x-rays. [13] and [14] both implemented similar classification using CNN while utilising transfer learning, classifying chest x-ray images into normal, viral pneumonia and COVID-19 cases. [15] utilised CNN as a fixed feature extractor, and used a support vector machine (SVM) for classification.

The usage of CNN contributed to the performance of these lung disease detection. CNN is very suitable for image classification and object detection because it automatically extracts and learns features, discarding the necessity for manually extracting features. The usage of deep learning can speed up the disease identification process [16].

In this paper, we demonstrate that it is possible to detect COVID-19 and pneumonia by only using edge images. Edge images were considered because edges represent the shape feature of an image, and shape is an important and powerful feature used for image classification [17]. Previous studies also showed that lungs with diseases produced unusual edges compared to normal lungs[18][19].

3. Literature Review

In this section, the enhancement of Canny edge detection is explained. Subsequently, to investigate the benefit of our edge detection method, we applied the edge detection technique to several datasets which will later be used for COVID-19 and pneumonia detection.

3.1. Contrast Enhanced Canny Edge Detection

In this project, a method named Contrast Enhanced Canny Edge Detection (CEED-CANNY) was proposed. This method combines local morphological contrast enhancement and Canny edge detection.

First, morphological contrast enhancement was performed on a grayscale image, subsequently, Canny edge detection was applied. In morphological contrast enhancement, a filter is applied. If the central pixel is closer to the local maximum, this filter replaces it with the local maximum. Else, the central pixel is replaced by the local minimum. Afterwards, Canny edge detection was implemented to identify edges existing in the x-rays. Figure 3 displays the pseudocode for implementing the Contrast Enhanced Canny Edge Detection technique.

1. Setup

$$Original\ pixel\ value: P_o$$
$$Local\ maximum: P_{max}$$
$$Local\ minimum: P_{min}$$

2. Morphological contrast enhancement

$$D_1 = P_o - P_{max}$$
$$D_2 = P_o - P_{min}$$
$$if\ D_1 < D_2 : P_o = P_{max}$$
$$else : P_o = P_{min}$$

3. Noise Reduction using Gaussian smoothing

$$Original\ image: I(i,j)$$
$$Smoothed\ image: F(i,j) = G \times I(i,j)$$
$$Gaussian\ function: G(i,j) = \frac{1}{\sqrt{2\pi\sigma^2}} exp - (\frac{i^2 + j^2}{2\sigma^2})$$

4. Finding Intensity Gradient of the Image

$$magnitude\ of\ the\ edge: G = \sqrt{G_x^2 + G_y^2}$$
$$direction\ of\ the\ gradient: \theta = \tan^{-1}(\frac{G_y}{G_x})$$

5. Non-maximum Suppression
6. Hysteresis Thresholding

$$if\ G < t_{low} : discard\ the\ edge$$
$$if\ G > t_{high} : discard\ the\ edge$$
$$else : keep\ the\ edge$$

Figure 3. Pseudocode for Contrast Enhanced Canny Edge Detection.

3.2. Application of the Edge Detection technique for COVID-19 and Pneumonia Detection in Digital X-Ray Images Using CNN

To conduct COVID-19 and Pneumonia Detection on Edge X-Ray using CNN, first, datasets were obtained from various sources. The datasets used were: Joseph Cohen's GitHub repository [20], Andrew's Kaggle dataset (available at https://www.kaggle.com/andrewmvd/convid19-X-rays) and Kermany's dataset [21]. These datasets contained x-ray images with lung conditions of COVID-19, pneumonia, and normal (healthy). 196 images from each lung condition were collected. Thus, a total of 588 images were used in this experiment.

Next, Contrast Enhanced Canny Edge Detection was applied to all the images, producing x-ray images with only the edges. This will be known as the Edge dataset.

Afterwards, the Edge dataset was divided into two separate sets. 90 percent of the images were put in a training set, while the remaining 10 percent were put in a testing set. The training set was used for classifier training, while the testing set was used for testing.

For training, two CNN architectures, namely VGG16 [22] and InceptionV3 [23], were used to extract features from these images. Then, the features extracted will become the input for a new model. And this produces two classifiers, namely VGG16 classifier and InceptionV3 classifier.

For testing, the VGG16 classifier and InceptionV3 classifier were used to classify the images in the testing set. Their accuracy, sensitivity, and specificity were recorded. False positive (FP), false negative (FN), true positive (TP) and true negative (TN) were recorded to determine accuracy, sensitivity, and specificity.

True positive rate, or more commonly known as sensitivity, is the likelihood that a test result will be positive when the disease is present.

$$Sensitivity(\%) = \frac{TP}{TP+FN} \times 100 \quad (3)$$

True negative rate, or more commonly known as specificity, is the likelihood that a test result will be negative when the disease is absent.

IOP Conf. Series: Materials Science and Engineering **979** (2020) 012016 doi:10.1088/1757-899X/979/1/012016

$$Specificity(\%) = \frac{TN}{TN+FP} \times 100 \quad (4)$$

Accuracy is the overall probability that a patient will be correctly classified.

$$Acuracy(\%) = \frac{TP+TN}{TP+FN+FP+TN} \times 100 \quad (5)$$

4. Results and Discussion

For this research, two tests were conducted. The first one was the mean square error (MSE) test. This test was used to compare out edge detection method with the traditional Canny method. The second test was COVID-19 and Pneumonia Detection on Edge X-Ray Using CNN. This test was conducted to demonstrate that it is possible to detect COVID-19 and pneumonia by only using edge images.

4.1. MSE Test

In this section, we compare our CEED-CANNY with the traditional Canny technique. To compare their performances, MSE test was performed. The MSE permits the calculation of the "true" pixel values of the actual image to the noisy image. The average of the squares of the "error" between the original image and the degraded image are signified by the MSE. The error here refers to the difference between the values of the actual image and the noisy image. A lower MSE means that the image is produced with a better quality. The MSE is represented in the following mathematical equation:

$$MSE = \frac{1}{mn}\sum_0^{m-1}\sum_0^{n-1}\|f(i,j) - g(i,j)\|^2 \quad (6)$$

where:

- *m* is the number of rows in the image, while *i* is the row index
- *n* is the number of columns in the image, while *j* is the column index
- *f* is the matrix of the actual image
- *g* is the matrix of the noisy image in question
- *MAXf* is the maximum signal value that in the actual image

Five chest x-ray images were selected as samples. These images were obtained from Kermany's dataset and Cohen's dataset. Figure 4 to Figure 8 show the output images of Canny and CEED-Canny respectively.

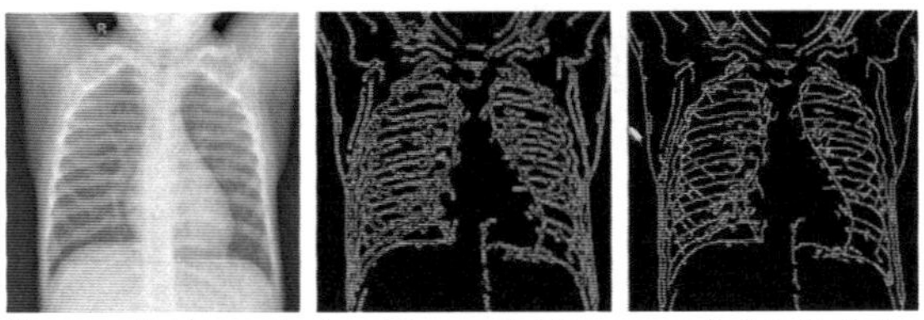

Figure 4. Sample 1 – Normal (Kermany) (a) Original Image (b) Canny (c) CEED-Canny.

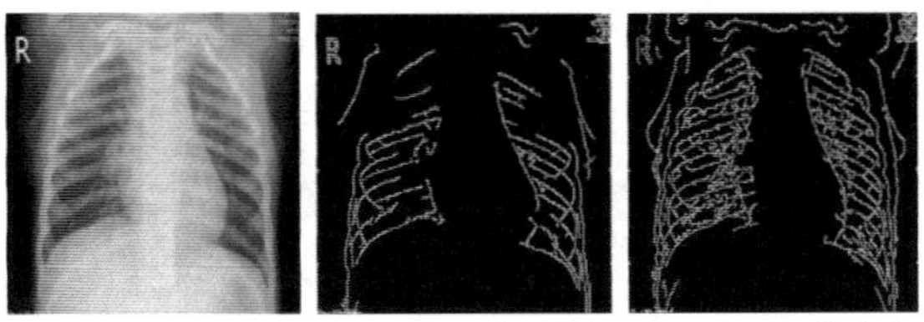

Figure 5. Sample 2 – Pneumonia (Kermany) (a) Original Image (b) Canny (c) CEED-Canny.

IOP Conf. Series: Materials Science and Engineering **979** (2020) 012016 doi:10.1088/1757-899X/979/1/012016

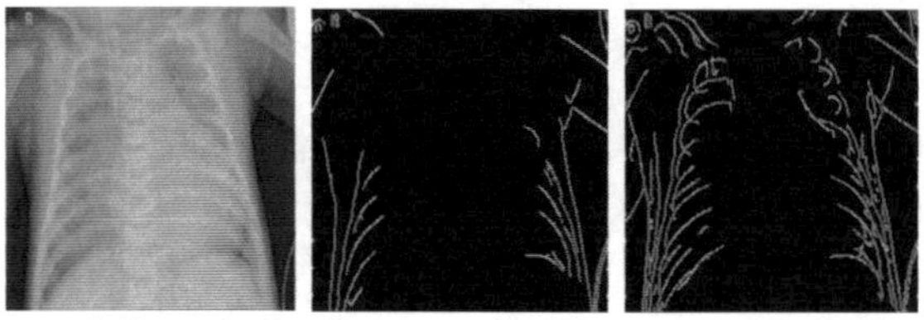

Figure 6. Sample 3 – Pneumonia (Kermany) (a) Original Image (b) Canny (c) CEED-Canny.

Figure 7. Sample 4 – COVID-19 (Cohen) (a) Original Image (b) Canny (c) CEED-Canny.

Figure 8. Sample 5 – COVID-19 (Cohen) (a) Original Image (b) Canny (c) CEED-Canny.

Both techniques were able to detect edges remarkably. From our observation, the CEED-Canny method produced more edges. Table I shows the result of the MSE test.

TABLE 1. MSE test result for Canny and CEED-Canny Techniques.

Sample	CANNY MSE	CEED-CANNY MSE
1. Normal (Kermany)	0.1192474365234375	0.11898040771484375
2. Pneumonia (Kermany)	0.11791610717773438	0.11594390869140625
3. Pneumonia (Kermany)	0.12229156494140625	0.12009811401367188
4. COVID-19 (Cohen)	0.12252044677734375	0.1218414306640625
5. COVID-19 (Cohen)	0.12311553955078125	0.12115478515625

Results show CEED-Canny gives lower MSE values, while Canny gives higher MSE values. This signifies that CEED-Canny produces edge images that are more accurate than the traditional Canny.

4.2. COVID-19 and Pneumonia Detection on Edge X-Ray Using CNN

To investigate the benefit of our edge detection method, we applied the edge detection technique to produce an alternate dataset for COVID-19 detection. After the test was conducted, the results were taken and recorded. Table 2 presents the sensitivity, specificity, and accuracy of both CNN architectures.

TABLE 2. Results of COVID-19 and pneumonia on edge x-ray using CNN.

Model	ACCURACY	COVID-19		Normal		Pneumonia	
		Sensitivity	Specificity	Sensitivity	Specificity	Sensitivity	Specificity
VGG16 Classifier	0.9369	0.9684	0.9746	0.9474	0.9625	0.8948	0.9657
InceptionV3 Classifier	0.9035	0.9526	0.9579	0.9158	0.9379	0.8421	0.9543

Based on the result, it is established that the classifiers were able to categorize edge images into COVID-19, normal, and pneumonia cases. Not only that, but the accuracy, sensitivity, and specificity were quite high.

5. Conclusion

This paper presents a modified Canny edge detection method. This method, named CEED-Canny, was the combination of local morphological contrast enhancement and Canny edge detection. The MSE values for the proposed method was lower than that of the traditional Canny method. Furthermore, this paper also showed that edge images can be used to train COVID-19 and pneumonia detection classifiers with high accuracy, sensitivity and specificity.

There are some limitations to this study. First, even though CEED-Canny performed better when compared to the traditional Canny method, no comparison was made with other edge detection methods. This is because not all articles that proposed edge detection methods include source codes, and the sample images used are different. Future studies may perform a survey of all the recent edge detection method, and come up with a standardized test. Second, this lung disease detection approach was only tested on a relatively small dataset. In the future, more data should be gathered for training and testing, as this will most likely improve the classification accuracy and generalizability.

Acknowledgement

The work presented is funded by Universiti Malaysia Sabah (UMS) through grant SDK0191-2020.

References

[1] Kumar I., Rawat J. and Bhadauria H. S., "A Conventional Study of Edge Detection Technique in Digital Image Processing," Int. J. Comput. Sci. Mob. Comput., vol. 3, no. 4, pp. 328–334, 2014.

[2] Rashmi, M. Kumar, and Saxena R., "Algorithm and Technique on Various Edge Detection: A Survey," Int. J. Signal Image Process., vol. 4, no. 3, pp. 65–75, 2013.

[3] Makandar A. and Halalli B., "Image Enhancement Techniques using Highpass and Lowpass Filters," Int. J. Comput. Appl., vol. 109, no. 14, pp. 12–15, 2015.

[4] Canny J., "A Computational Approach to Edge Detection," IEEE Trans. Pattern Anal. Mach. Intell., vol. 8, no. 6, p. pp.679-698, 1986.

[5] Lahani J., Bade A., Sulaiman H. A. and Muniandy R. K., "InTEC : Integration of Enhanced Entropy – Canny Technique for Edge Detection in Digital X-Ray Images," ASM Sci. J., vol. 11, no. 3, pp. 161–167, 2018.

[6] Webb A., Introduction To Biomedical Imaging. John Wiley & Sons, Inc., Hoboken, New Jersey, 2003.

[7] Ai T. et al., "Correlation of Chest CT and RT-PCR Testing in Coronavirus Disease 2019 (COVID-19) in China: A Report of 1014 Cases," Radiology, p. 200642, Feb. 2020.

[8] Waheed A., Goyal M., Gupta D., Khanna A., Al-Turjman F. and Pinheiro P. R., "CovidGAN: Data Augmentation Using Auxiliary Classifier GAN for Improved Covid-19 Detection," IEEE Access, vol. 8, pp. 91916–91923, 2020.

[9] Labhane G., Pansare R., Maheshwari S., Tiwari R. and Shukla A., "Detection of Pediatric Pneumonia from Chest X-Ray Images using CNN and Transfer Learning," Proc. 3rd Int. Conf.

Emerg. Technol. Comput. Eng. Mach. Learn. Internet Things, ICETCE 2020, no. February, pp. 85–92, 2020.

[10] Li L. et al., "Using Artificial Intelligence to Detect COVID-19 and Community-acquired Pneumonia Based on Pulmonary CT: Evaluation of the Diagnostic Accuracy," Radiology, vol. 296, no. 2, pp. 65–71, 2020.

[11] Butt C., Gill J., Chun D. and Babu B. A., "Deep learning system to screen coronavirus disease 2019 pneumonia," Appl. Intell., 2020.

[12] Wang L., Lin Z. Q. and Wong A., "COVID-Net: A Tailored Deep Convolutional Neural Network Design for Detection of COVID-19 Cases from Chest X-Ray Images," arXiv, 2020.

[13] Apostolopoulos I. D. and Mpesiana T. A., "Covid - 19: automatic detection from X - ray images utilizing transfer learning with convolutional neural networks," Phys. Eng. Sci. Med., pp. 1–6, 2020.

[14] Narin A., Kaya C. and Pamuk Z., "Automatic Detection of Coronavirus Disease (COVID-19) Using X-ray Images and Deep Convolutional Neural Networks," arXiv e-print, 2020.

[15] Sethy P. K., Behera S. K., Ratha P. K. and Biswas P., "Detection of coronavirus disease (COVID-19) based on deep features and support vector machine," Int. J. Math. Eng. Manag. Sci., vol. 5, no. 4, pp. 643–651, 2020.

[16] Rahaman M. M. et al., "Identification of COVID-19 samples from chest X-Ray images using deep learning: A comparison of transfer learning approaches," J. Xray. Sci. Technol., vol. 28, no. 5, pp. 821–839, 2020.

[17] Ramamurthy B. and Chandran K. R., "Content based Image Retrieval for Medical Images using Canny Edge Detection Algorithm," Int. J. Comput. Appl., vol. 17, no. 6, pp. 32–37, 2011.

[18] Kieu S. T. H., Hijazi M. H. A., Bade A. and Saffree Jeffree M., "Tuberculosis detection using deep learning and contrast-enhanced canny edge detected x-ray images," IAES Int. J. Artif. Intell., vol. 9, no. 4, 2020.

[19] Kieu S. T. H., Hijazi M. H. A., Bade A., Yaakob R. and Jeffree S., "Ensemble deep learning for tuberculosis detection using chest X-Ray and canny edge detected images," IAES Int. J. Artif. Intell., vol. 8, no. 4, pp. 429–435, 2019.

[20] Cohen J. P., Morrison P. and Dao L., "COVID-19 Image Data Collection," arXiv, 2020.

[21] Kermany D. S., Goldbaum M., Cai W. and Lewis M. A., "Identifying Medical Diagnoses and Treatable Diseases by Image-Based Deep Learning," Cell, vol. 172, no. 5, pp. 1122-1131.e9, 2018.

[22] Simonyan K. and Zisserman A., "Very Deep Convolutional Networks For Large-Scale Image Recognition," arXiv: 1409.1556, pp. 1–14, 2015.

[23] Szegedy C., Vanhoucke V., Ioffe S., Shlens J. and Wojna Z., "Rethinking the Inception Architecture for Computer Vision," in The IEEE Conference on Computer

IOP Conf. Series: Materials Science and Engineering **979** (2020) 012017 doi:10.1088/1757-899X/979/1/012017

Mobile based augmented reality for flexible human height estimation using touch and motion gesture interaction

N A Ismail *, C W Tan, S E Mohamed, M S Salam and F A Ghaleb

School of Computing, Faculty of Engineering, Universiti Teknologi Malaysia, 81310 UTM Skudai, Johor, Malaysia.

*azman@utm.my

Abstract. Human height measurement can be achieved by using contact or non-contact techniques. Contact technique is the traditional measuring method which required human resources to perform the measurement. In contrast, for non-contact technique, several kinds of research for measurement have been conducted, mostly with image-processing methods and only a few with the Augmented Reality (AR) approach. The current measuring approaches mostly required external hardware such as laser pointer or artificial fiducial such as 2D markers. In this paper, the world tracking technique and Visual Inertial Odometry is the method used to estimate the human height. The main aim of this paper is to accurately estimate the human height using augmented reality (non-contacted measurements). The methodology used the Apple ARKit plugin, which is the software development tools to build an augmented reality application for IOS device. An algorithm was designed by using Golden Ratio rules to estimate human height from the lower part of human knee; The estimation result is displayed using AR technology to allow the justification of the accuracy of the result. The application is tested with four different measuring methods. The normal full-height measurement result had a 1.13cm (0.73%) bias and a 1.34cm (0.88%) Root Mean Square Error (RMSE); the self-full height measurement had a result of 0.89cm (0.58%) bias and a 1.27cm (0.83%) RMSE; the normal height estimation from the lower part of knee measurement had a result of 0.12cm (0.06%) bias and a 1.34cm (0.89%) RMSE; the self-height estimation from the lower part of knee measurement had a result of 0.15cm (0.09%) bias and a 1.04cm (0.66%) RMSE. The results show that the mobile phone with VIO can be a potential tool for obtaining accurate measurements of human height.

1. Introduction

Commonly, traditional measurements can only be made with the use of measuring tools such as measuring tape or ruler. Although some researches using different measuring approaches such as image-processing methods and marker-based Augmented Reality approach had been done, it was also limited to the help of external sources such as laser pointer or 2D marker [4, 11, 15].

The camera in mobile devices nowadays is equipped with inertial sensors such as 3-axis accelerometer and 3-axis gyroscope. It gave the device 6-axis motion sensing and rotation about the gravity, enabling a new level of spatial awareness in measurement applications.

The term, Visual Inertial Odometry (VIO) arises due to the widely used of the combination of visual and inertial measurement in the robotic field. As mentioned by Rebecq [9], Visual Inertial Odometry is the task of estimating a sensor's ego-motion from a combination of images and measurements from an Inertial Measurement Unit (IMU), which plays a significant role in numerous application fields such as augmented reality (AR) or virtual reality (VR) applications. Currently, well-known applications that utilize VIO in the mobile device are Apple ARKit and Google Project Tango. With VIO, more accurate tracking can be done. Thus, a new way of measuring distance, length or height can be achieved.

Instead of just taking full human height measurement, human height can also be measured by just measuring a particular part of the body. The divine proportion, Golden Ratio, which was originated by a Greek sculptor and mathematician, Phidias, is a ratio of distances in simple geometric figures [3]. There has been evidence that the human body proportion does follow the Golden Ratio rules and several measurement rules had been derived from that [5, 6].

In this paper, an application utilized the VIO in a mobile device using the built-in camera to obtain the human height using the AR technology for displaying result is proposed. The visual and inertial data are used for obtaining human height. The system tracks visual data of the ground in which people stand on and also track the inertial data of the camera movement. The designed algorithm is then used to compute height measurement using the visual and inertial data.

Furthermore, the Golden Ratio rule in human body proportion was also implemented to estimate the human height by just measuring the lower parts of the knee. This feature is quite useful to get the height estimation for those who have disability problem.

In the following section, the theory and related application are first presented, followed by the methodology, results, and lastly, the conclusion is described.

2. Backgound

2.1. Visual Inertial Odometry (VIO)

Visual Inertial Odometry is described as a technique to estimate the change in mobile position and orientation using the built-in camera and Inertial Measurement Unit (IMU) sensor [7]. According to Rebecq, VIO is the fusion of IMU in the Visual Odometry (VO) system which estimates the ego-motion (the 3D camera motion from a combination of images and the IMU's measurements) [9].

2.2. Visual Odometry

Visual odometry can be defined as the prediction of camera motions in sequence depending on the perceived pixels' movement in the sequence of images. There are four components in the visual odometry: camera calibration, the feature tracker, the rigid motion estimation algorithm, and the algorithm that matches the point of the features descriptions, the Random Sample Consensus (RANSAC) algorithm.

Camera calibration algorithm cannot be considered as a fragment of the visual odometry, due to it being only a one-time execution which is in the initialization process in which the determination of the transformation between the ground coordinate system (CS) and the camera CS occur [13, 14]. Hence, for the real-time functionality on the mobile device, three main essential parts of visual odometry must be applied, and they are the Kanade- Lucas-Tomasi (KLT) algorithm, the RANSAC algorithm, and the algorithm to determine the rigid transformation.

KLT algorithm [8, 10, 12] is a feature tracker used to find out the movement between two subsequent captured images at time t and $t + \Delta t$.

The RANSAC algorithm is an iterative scheme that used a set of data observed, including the outliers to predict the parameters of a mathematical model [13, 14]. By using this algorithm, it allows robust estimation of the parameter of a model which suits the given data even when the data contain elements highly different from the precise values with the limitation of up to 50% elements.

For the algorithm that determines the rigid motion, only a single point movement that can indicate the whole points' movement is described, instead of describing all the individual points' movement to simplify the computational complexity.

2.3. Inertial Measurement Unit (IMU)

The IMU is an independent system which allows the measurement of linear and angular motion by using multiple sensors such as accelerometer, gyroscope and sometimes magnetometer. With an accelerometer, the device's acceleration, tilt and vibration can be detected to identify the movement and orientation. The gyroscope allows the identification of the relative directions: up, down, left, right, forward and backwards. Rotation around the three axes is also part of the gyroscope capability. The axis orientation is not affected by tilting of the mounting. Thus, the gyroscope is used to help provide consistency or to preserve a reference direction. Next is the magnetometer, which allows the detection of magnetic north and is most commonly used in GPS to determine the user's location.

With the capability of determining the direction of gravity (which is in line with the Z-axis of the world CS) by the accelerometer and the direction of magnetic north (Y-axis) provided by a magnetometer, X-axis can be obtained by the computing the cross product of these two directions. However, a gyroscope can measure angular velocity (relative rotation) accurately and its high responsiveness (high sampling rate). It is needed for a more accurate and stable measurement. It can prevented a false rotation returned by the magnetometer in the presence of magnetic interference.

The occurrence of noise, drift and bias is often present in these inertial sensor's measurement, so they are seldom used individually. However, these biases can be highly eliminated through the combination of all three sensors into one sensor, namely the Inertial Measurement Unit (IMU). This combination is typically done by using the Kalman filter algorithm as it is common sensor fusion and data fusion algorithm. The standard method used to measure device movement in between IMU readings is dead reckoning [1]. It is a process of computing current position by using a previously fixed or determined position and advancing that position based upon known or estimated speeds over the time elapsed. The problem is the cumulative error, where the error accumulates over time due to noises and biases in IMU data [1].

2.4. Golden Ratio on Human Body

Golden Ratio is simply a proportion of two unequal segments at which the proportion of the long segment to the short segment is equal to the total of the two segments to the long segment (1:1.61803398874989) [6]. As shown in Figure 1, the proportion of the longer segment a to the short segment b (a:b) is equal to the total of a and b to a. The value of φ is given with the number as 1.61803398874989.

Leonardo da Vinci had established 15 proportion rules [5, 6] used to human modelling based on his work on Vitruvian Man around 1490, as shown in Equation 1.

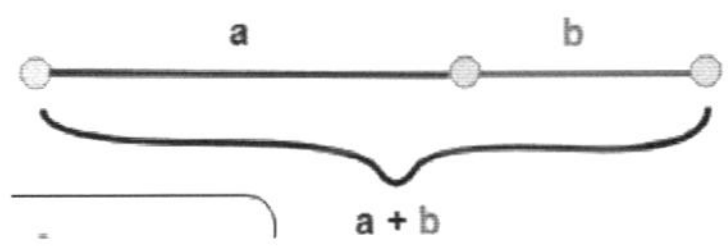

Figure 1. Golden Ratio representation in the line segment.

$$\frac{a+b}{a} = \frac{a}{b} = \varphi \tag{1}$$

Further work with golden ratio proportions on the human body had been done by a German engineer, Ernst Neufert in 1936 with the book named "Bauentwurfslehre" [5, 6]. Figure 3 shows the 29 proportional measurements derived for the human model, where some of the rules were used in this project for human height estimation.

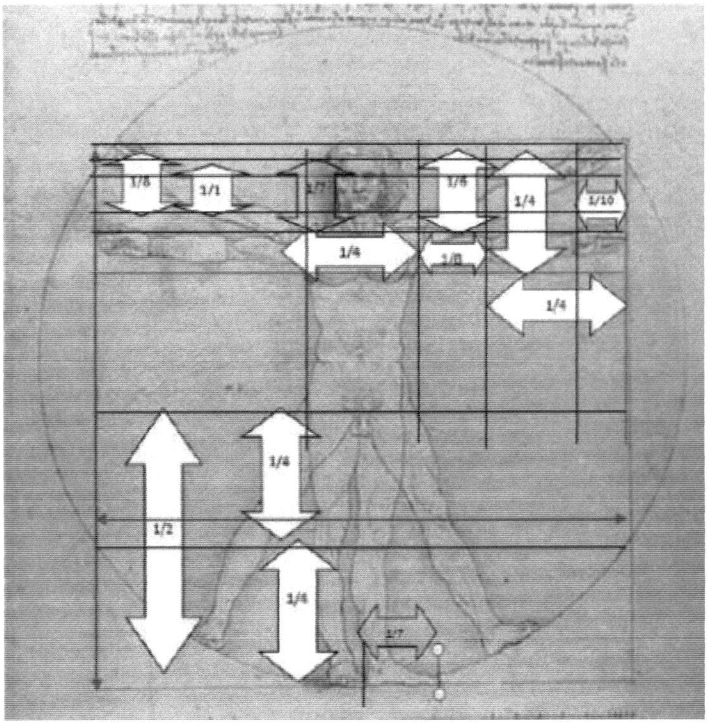

Figure 2. Vitruvian man with proportions by Leonardo da Vinci [5, 6].

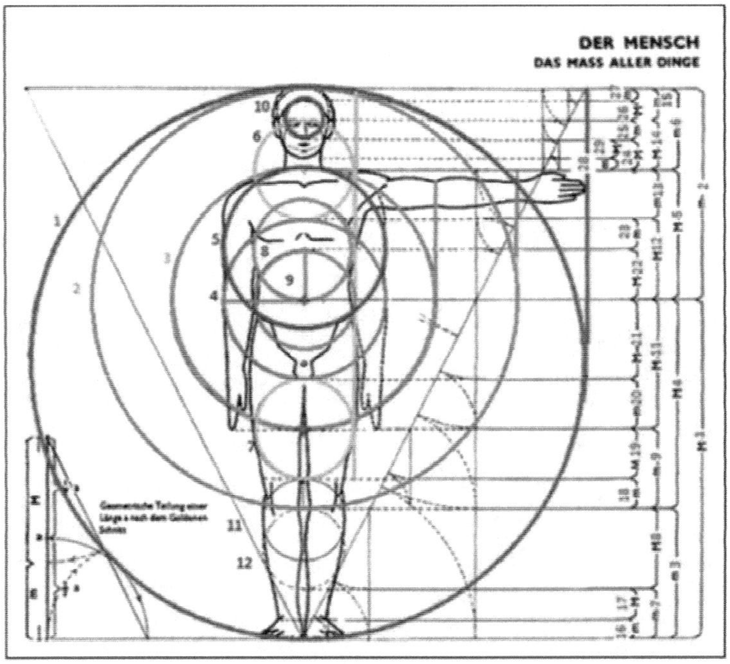

Figure 3. Bauentwurfslehre by Ernst Neufert. [5, 6].

2.5. Related Work

Guan et al [16] have proposed a method using single image containing only a human face to estimate body height in 2009. Eyes, lips and the chin are extracts to identity the height. The height estimation made was based on the facial vertical golden proportion.

Wang and Chen did the project in 2012, which utilized a laser pointer and a camera associated with the image processing technique to get the human height measurement. To perform measurement operation, the user just need to stand on the measurement mechanism, and the detection device on top will automatically calculate the height by using the triangular distance measurement principle [15].

Another image processing research of estimating people height had been done in 2018 by Petr, Kamil, et al. In order to determine the height, a full human image which was captured by a single camera, as well as the presence of at least one object with known size in the image, are required. Several images processing technique had been used such as camera calibration, edge line detection, Hough transform and many more. [17].

There was also research on using the square template to identify the human height performed by a grouped of four researchers in Srinakharinwirot University in 2018. Four steps are used to get the height estimation which are image acquisition, template matching, background subtraction, and lastly estimating the height based on pixel size [18].

Research on estimating human height with object (foot) size had been done at the University of Pennsylvania, in 2018. The human foot was extracted from the captured image using colour mask segmentation to separate the noise of the background and foot. A coin was used as the reference for depth estimation. A linear regression model is trained and used to predict the result [19].

An AppyKids Toy Box application called "How Tall Are You" had been created by Growl Media in 2018 for the purposed to track and measure children height. The application also provided picture saving and sharing functions. The system needs to analyze the room before it started to do the measurement. After that, the user needs to keep the device at the level of the foot and tap the screen. An animated giraffe will be displayed on the screen. The device is then raised to the head level and tap again to complete the measurement [2].

Therefore, this research will fill the gap by applying the golden ratio in estimating human height in the AR environment, which will contribute to assisting disabled people in measuring their height.

3. Proposed Method

3.1. System architecture

The description of the system for this application is shown in Figure 4. Unity 3D and ARKit plugin were used to develop this application, while XCode was used to build and run this application onto the device. The compatible devices need to be IOS 11.0 and above with camera, gyroscope, and accelerometer.

For the application layer, the camera's visual tracking will first identify the surface by getting the features to point data from the plane detection function provided by ARKit plugin. Once the plane surface data is obtained, the application will proceed to get the hit-testing result from the centre screen in the form of 3D world coordinates once the user clicks on the add point button. Once the starting point is set, the application will utilize the device inertial sensors data to update the ending point.

Once the add button has been pressed again by user, the system will get the current measurement state to identify which ending point to be used. If the state is self-measurement, then the ending point will be based on the camera position; otherwise, if the state is a normal measurement, the point will be based on the pointed position by the centre of the screen. There are two different algorithms to compute the height, which is the normal measure algorithm and golden ratio estimate algorithm. Once the height had been calculated, it was displayed together with the line from the starting point to the ending point using the AR technology.

Figure 5 shows full human height was taken as measurement where it was computed using the normal measure algorithm explained in the next subsection as in Equations (2) and (3). In contrast, Figure 6 shows only the lower part of the knee was taken as the measurement using a normal measure algorithm.

The human height can be obtained through the design golden ratio estimate algorithm as in Equation (5).

3.2. Estimation of Height Increment

The total height is updated with the motion of the mobile device once the starting point had been set. Figure 7 shows the illustrated diagram involved in computing the height increment per frame updated with device motion. The start point, p1 is the point on the ground surface that is obtained from plane detection. Camera position can be obtained through the tracking of the device movement.

Both camera position and the start point, p1 obtained are based on the world origin of the AR session. To get the height of the camera above ground (h1), assumption needs to be made on the position of the point pCam, where pCam is parallel to y-axis of p1. The height h1 can then be obtained by calculating the distance from the camera to pCam.

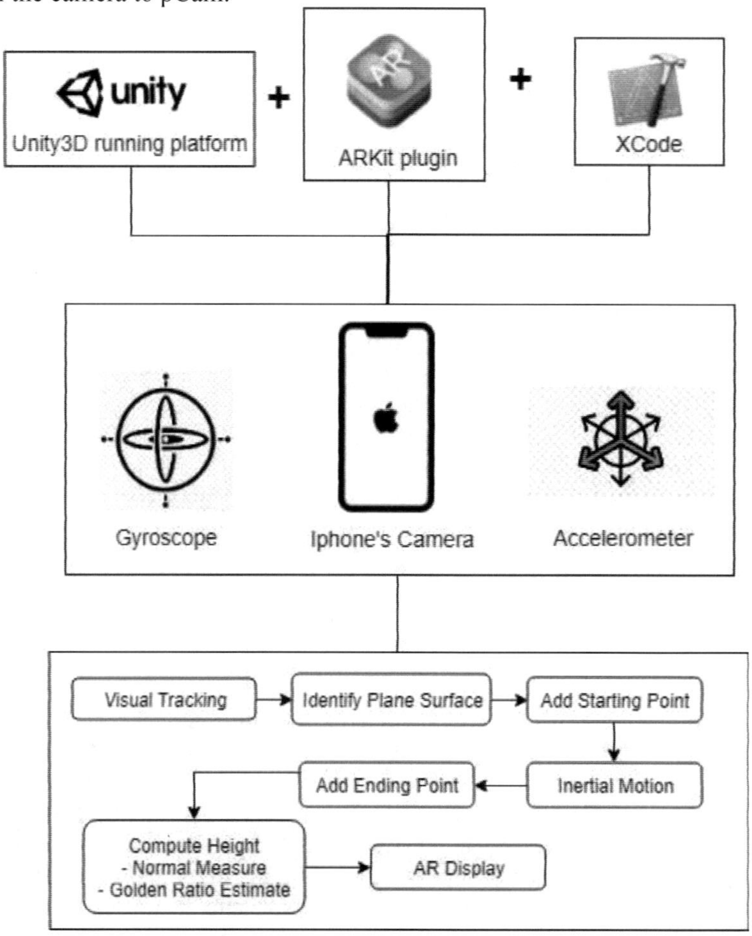

Figure 4. System Architecture.

The angle of elevation, θ, is obtained from the orientation of the device. The overall equation used to obtain the height increment is in Equations (2) and (3) below.

$$h2 = \tan(\theta) \times dis\tan ce\, to\, p1 \tag{2}$$

$$Height = h1 + h2 \tag{3}$$

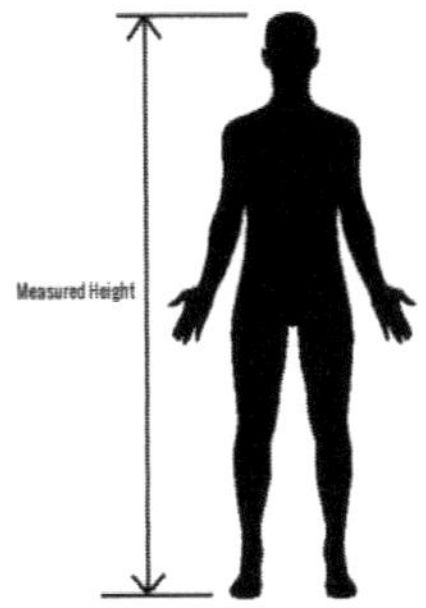

Figure 5. Full height measurement.

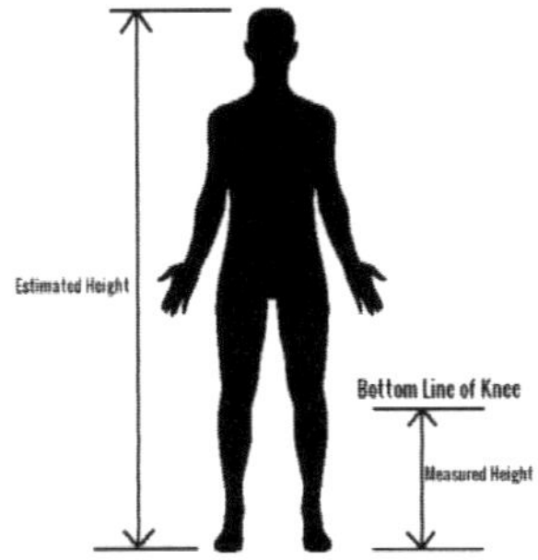

Figure 6. Lower part of knee measurement to estimate human height.

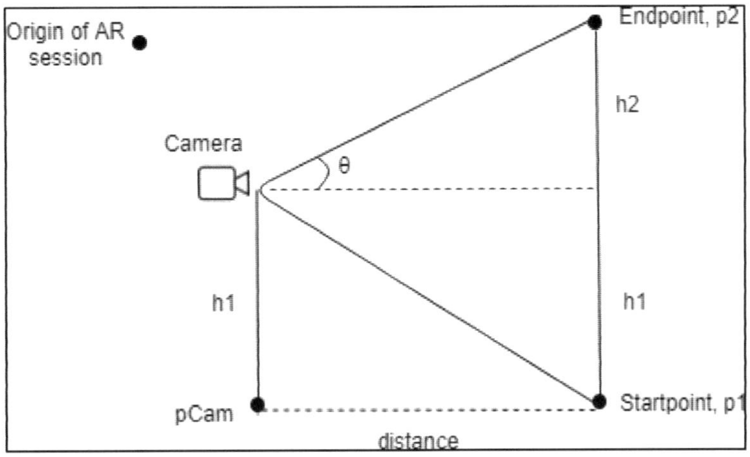

Figure 7. Height Computation Schema.

3.3. Golden Ratio Height Estimate

Since the human body followed the golden ratio rules as described in Figure 2, human height can be obtained by measuring just a small part of our body. The human height was estimated with a known height from foot to the bottom line of the knee. The algorithm involved the use of M3, M4, M1 and M2 as illustrated in Figure 3.

M3 – The height from foot to the bottom line of knee
M4 – The height from the bottom line of the knee to the navel
M1 – The height from foot to the navel
M2 – The height from the navel to the top of the head

Generally, the value of M3 is estimated using the rules and method in Figures 1 and 3. With known value M3, M4 can be obtained by using the Equation (4) obtained from the golden ratio rule where the value φ is 1.61803398874989.

$$M4 = M3 \times \varphi \tag{4}$$

The height from foot to the navel, M1 is calculated where M1 = M3 + M4. Next, the height from the navel to the top of the head, M2 is also calculated with the equation derived from the golden ratio proportion as in Equation (5).

IOP Conf. Series: Materials Science and Engineering **979** (2020) 012017 doi:10.1088/1757-899X/979/1/012017

$$M2 = M1 \div \varphi \tag{5}$$

Lastly, the total height is obtained by calculating the sum of M2 and M1. The overall equation used for estimating the human height by measuring the lower part of human knee is shown in Equation (6).

$$Height_{Est} = M3 \times (1 + \varphi) \times \varphi \tag{6}$$

3.4. Evaluation of the Accuracy of Human Height Estimation

The accuracy of the human height estimations was evaluated using bias, relative bias (relBIAS), root mean square error (RMSE), and relative RMSE (relRMSE), as shown in the Equations (7), (8), (9) and (10) below.

$$\text{BIAS} = \frac{\sum_{i=1}^{n}(x_i - x_{iref})}{n} \tag{7}$$

$$\text{relBIAS} = \frac{\sum_{i=1}^{n}(x_i - x_{iref})}{n} \times 100\% \tag{8}$$

$$\text{RMSE} = \sqrt{\frac{\sum_{i=1}^{n}(x_i - x_{iref})^2}{n}} \tag{9}$$

$$\text{relRMSE} = \sqrt{\frac{\sum_{i=1}^{n}(x_i / x_{iref} - 1)^2}{n}} \times 100\% \tag{10}$$

Based on the equations above, x_i is the ith measurement, x_{iref} is the ith actual value, while n is the number of trials or estimations. Since RMSE is a measure of how well the model performed by measuring the difference between estimated values and the actual values, it is chosen to be the method for evaluating the accuracy.

4. Results

The results of the project are analyzed to evaluate the accuracy of the proposed technique, two measuring methods can be made with this application, which is a normal measuring method and self-measuring method. The purpose of normal measuring method is to help people measuring the height while the self-measuring method is to measure self-height. For each measuring method, it is split into two categories which are full height measurement and height estimation by measuring the only lower part of the knee. This section first showed some screenshot output for height measurement using both algorithms (golden ratio estimation and normal measurement), followed by presenting the evaluation results on human height estimation based on all measuring methods implemented in this application.

4.1. Output Result

Figure 8 showed the screenshot of example height measurement result for full height measurement using only the normal measure algorithm as in Equations (1) and (2). The measurement was taken from the ground until the head level—the two white dots indicating the start and the endpoint of the measurement. The height text was displayed in the middle of the line rendered as in Figure 8.

Figures 9 and 10 showed the example height estimation screenshot taken after measuring the lower part of the knee using the Equations (1) and (2). The final height was computed by using the golden ratio estimate algorithm as in Equation (5).

4.2. Evaluation on Human Height Estimation

The estimated results of the human heights using all four methods were similar to the actual human height measured, as shown in Figure 11.

IOP Conf. Series: Materials Science and Engineering **979** (2020) 012017 doi:10.1088/1757-899X/979/1/012017

The statistical results are summarized in Table 1. It is shown that all the estimation methods had relatively small BIAS and RMSE. Normal full height measurement had BIAS (-1.13 cm, -0.73%) with RMSE (1.34 cm, 0.88%), while self-full height measurement had BIAS (-0.89 cm, -0.58%) with RMSE (1.27 cm, 0.83%). Normal height estimation from the lower part of knee measurement had BIAS (0.12 cm, 0.06%) and RMSE (1.34 cm, 0.89%) whereas the self- height estimation from the lower part of knee measurement had BIAS (-0.15 cm, -0.09%) and RMSE (1.04 cm, 0.66%). The normal height estimation from knee measurement had the greatest dispersion of estimated values, as shown in Figure 12.

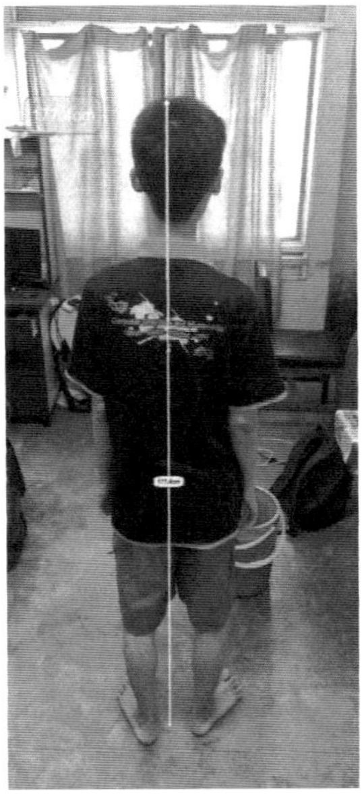

Figure 8. Full Height Measurement.

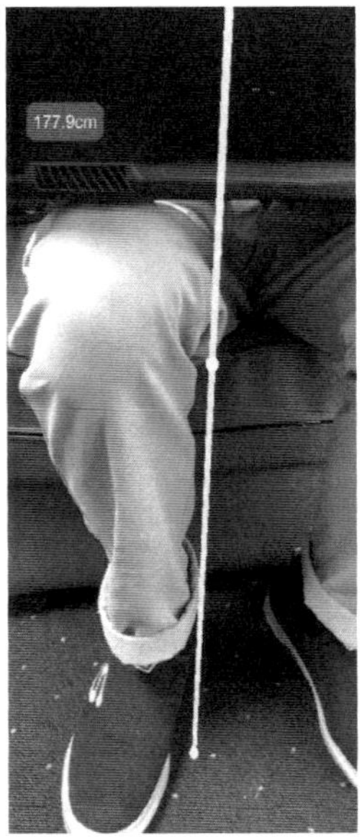

Figure 9. Height estimation result from only lower part of knee measurement.

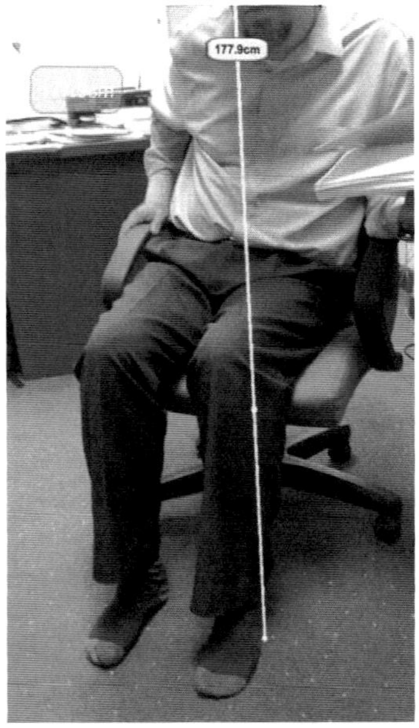

Figure 10. Height estimation result from only lower part of knee measurement.

Table 1. Accuracy of human height estimation using all available methods.

Methods	BIAS (cm)	relBIAS (%)	RMSE (cm)	relRMSE (cm)
Normal full height measurement (Normal)	1.34	0.88%	-1.13	-0.73%
Self-full height measurement (Self)	1.27	0.83%	-0.89	-0.58%
Normal height estimation from lower part of knee measurement (Normal Estimate)	1.34	0.89%	0.12	0.06%
Self-height estimation from lower part of knee measurement (self Estimate)	1.04	0.66%	-0.15	-0.09%

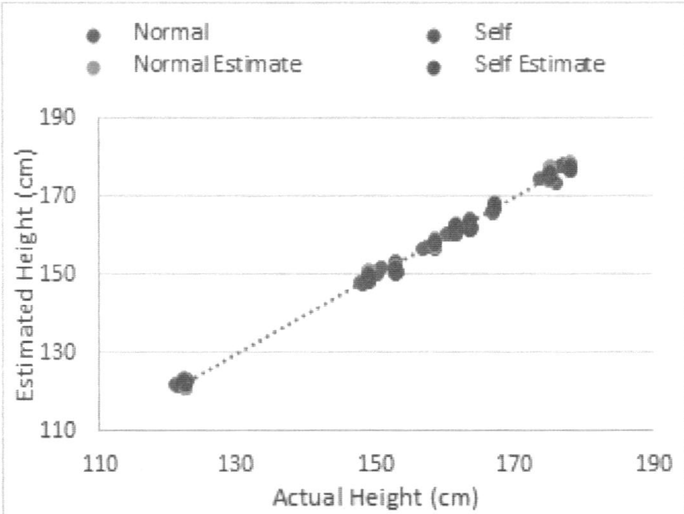

Figure 11. Scatter plot of human heights for all four available methods.

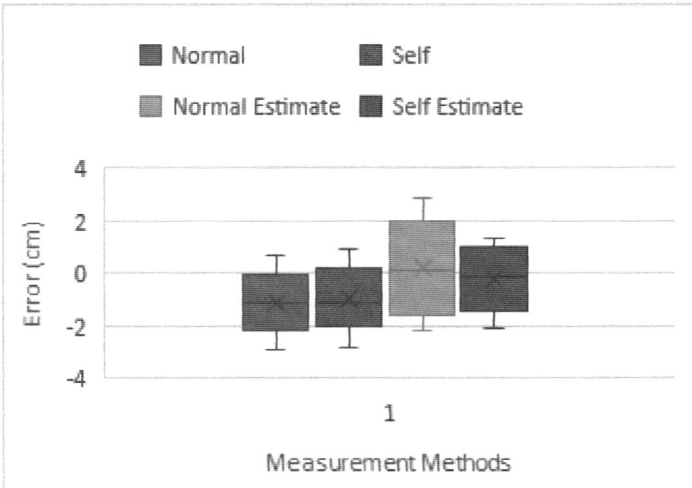

Figure 12. The errors in the human height estimations using different methods.

Based on the results of all the measuring methods, the normal measuring method has the highest RMSE (1.34 cm), while self-height estimation from knee measurement has the lowest RMSE (1.04 cm). It is believed that the reason leads to this difference in error might be the parallax error, where the viewing angle makes the line looks like a parallel to the targeted ending point, but in fact, it does not.

4.3. Comparison with other methods
A comparison with other methods was made as there were many papers studied on human height measurement. Table 2 shows the features comparison among other's methods and proposed method. The object size measurement method is a method that measures foot size to estimate human height and required a reference object which is a coin. For the proposed method and Wang and Chen method, both do not require any reference object. Full body image is captured for Wang and Chen method. The proposed method supports the measurement of the lower part of the knee to estimate the human height as well as a full body.

Table 2. Features comparison among other's methods and proposed method.

	Object size measurement method [19]	Wang and Chen Method [15]	Proposed Method
Reference Object	Yes (coin)	No	No
Full Body Captured	No. Only foot	Yes	No. Can be full body or lower part of knee
Type	Image Processing	Image Processing	Augmented Reality
Involve Golden Ratio	Yes	No	Yes

5. Conclusion
This paper presented a description of the application developed to estimate human height using mobile phone with VIO. The human height was estimated by full measurement or lower part of knee measurement. The results showed that this method had the potential to be used for accurate human height estimation.

It is recommended that further improvement can be made by applying automatic identification of the ending point for measurements such as the top of the head level, and the lower part of knee measurement.

Acknowledgement
The authors wish to thank the Ministry of Higher Education, Malaysia for funding this study under the Fundamental Research Grant Scheme (FRGS/1/2019/ICT04/UTM/02/8).

References
[1] Why is ARKit better than the alternatives? – 6D.ai – Medium (2017), https://medium.com/6dai/why-is-arkit-better-than-the-alternatives- af8871889d6a
[2] Use the AppyKids new augmented reality measuring tool to track your child's height! In: AppyKids (2018), https://www.appykids.com/loved- beyond-measure-track-and-save-your-childs-height-using-augmented- reality/
[3] Golden Ratio. In: from Wolfram MathWorld (2019), http://mathworld. wolfram.com/GoldenRatio.html
[4] Abdul-Adheem, W.: Distance Measurement using Dual Laser Source and Image (January 2012), https://www.researchgate.net/publication/309716162_Distance_Measurement_using_Dual_Laser_Source_and_Image_Processing_Techniques
[5] E, A.T.: An Algorithm for Human Modeling in Information Technology Multimedia Using Human Biometrics Found in Golden Ratio, Vitruvian Man and Neufert. Fifth International Conference on e-Learning (econf) (2015)
[6] E, A.T., HS, A.B.: A Human Body Mathematical Model Biometric Using Golden Ratio: A New Algorithm. Machine Learning and Biometrics (2018)
[7] Gui, J., Gu, D., Wang, S., Hu, H.: A review of visual inertial odometry from filtering and optimisation perspectives. Advanced Robotics 29:1289–1301 (2015)
[8] Lucas, B.D., Kanade, T.: An Iterative Image Registration Technique with an Application to Stereo Vision. pp. 674–679. Morgan Kaufmann Publishers Inc. Canada (1981)
[9] Rebecq, H., Horstschaefer, T., Scaramuzza, D.: Real-time Visual-Inertial Odometry for Event Cameras using Keyframe-based Nonlinear Optimiza- tion (2017)
[10] Shi, J., undefined Tomasi: Good features to track. Proceedings of IEEE Conference on Computer

Vision and Pattern Recognition CVPR-94 (1994)

[11] Swaminathan, R., Schleicher, R., Burkard, S.: Happy Measure. Interna- tional Journal of Mobile Human Computer Interaction 5, 16–44 (2013)

[12] Tomasi, C., Kanade, T.: Detection and Tracking of Point Features. Inter- national Journal of Computer Vision (1991)

[13] Tomažič, S., Škrjanc, I.: Fusion of visual odometry and inertial navigation system on a smartphone. Computers in Industry 74:119–134. Computers in Industry 74, 119–134 (2015)

[14] Tomažič, S., Škrjanc, I.: Monocular Visual Odometry on a Smartphone. IFAC-PapersOnLine 48:227–232. IFAC-PapersOnLine 48(10), 227–232 (2015)

[15] Wang, C.M., Chen, W.Y.: The human-height measurement scheme by using image processing techniques. 2012 International Conference on In- formation Security and Intelligent Control (2012)

[16] Y. Guan, "Unsupervised human height estimation from a single image", Journal of Biomedical Science and Engineering, vol. 2, no. 6, pp. 425-430, 2009.

[17] Petr Kriz, Kamil Riha, Malay Kishore Dutta, Anushikha Singh, Nechiti Marius, "Determination of a Person's Height from Images Using a Known Object Size", Telecommunications and Signal Processing (TSP) 2018 41st International Conference on, pp. 1-5, 2018

[18] Nattapol Suwatkittiwong, Theekapun Charoenpong, Chamaipom Sukjamsri, Taweechai Ouypornkochagom, "Human Height Measurement by Square Template", Electrical Engineering Congress (iEECON) 2018 International, pp. 1-4, 2018.

[19] Muhammad Ahmad Shahid, Muhammad Aksam Iftikhar, Zaheer Ahmad Gondal, Muhammad Adnan, Saima Rathore, "Object Size Measurement through Images: An Application to Measuring Human Foot Size", Frontiers of Information Technology (FIT) 2018 International Conference on, pp. 298-302, 2018

Interactive mobile games development for children to learn about money

M N H Zamri[1]*, N A F Ismail[1], A Ali[1], N M Suaib[1] and S Sadimon[1]

[1] ViCubeLab, School of Computing, Faculty of Engineering, University Teknologi Malaysia, 81310 Skudai, Johor Bahru, Malaysia.

* mnhafizzamri@gmail.com

Abstract. Money notes and coins are used for transactions and are essential in daily human life. Children need to be exposed to enough knowledge about the purposes of money notes, and the most important is to identify the money correctly. Children begin to expose to identifying the money at school in standard one. However, depending on formal education is not enough because the children easily get bored, and they did not fully engage with formal education. The use of mobile games for learning purposes is becoming increasingly popular because they are interactive and able to attract students. Therefore, this research aims to develop mobile games to identify and learn about money. Three main phases have been carried out, and the first phase is preliminary research on the learning approach for children and data collection in money Malaysia Ringgit. The second phase is to design and develop mobile games. Three mini-games have been proposed, and the first game is to identify the money notes and coins. The second game is to learn about counting the money and coins, and the third game is to learn to use the money for purchasing purposes. After these three mini-games have been developed in the second phase, the third phase is the evaluation phase. The usability result shows the users agree the mobile games help them to identify money, calculate the value of money, and to perform basic purchasing activities. Based on the user acceptance results show that the mobile game works more effectively with adult or parental guidance. Overall, this research has successfully produced a mobile game that can help the children to learn to identify, count, and use money.

1. Introduction

Humans are a social species; humans need each other for supply and service needs. This results in the creation of the barter system for it to be benefited by them. In the barter system, one item would be traded for other items, and it's involved between two or more parties. However, this system turns out to be ineffective [1].

Money is defined as a tool of exchange and to calculate the value of something [2]. In modern times, money plays an important role in our economy. This is because money is used for almost all daily things like buying groceries, making online payments, loan payments, transferring them to others, and so on. The proper knowledge of money usage is required to complete the tasks. Various social and economic problems can be experienced if people do not have sufficient knowledge and skills in money management [3].

Therefore, early education for children on the basics of money management is important because it will help them learn skills that will be beneficial as they grow up. This statement is supported by [3] because, in this day and age, children need to be given enough knowledge about money so that they can build good habits, skills, and attitudes in managing their money for the future. Based on the facts, children in Malaysia are exposed to formal education about money starting from first grade in a primary school in the subject of Mathematics. However, parents need to play an important role in money education as they are responsible for giving money to their children.

2. Problem Background

A study conducted by [3] shows that primary school students in Malaysia have a moderate understanding of money. For example, when shopping at the store or school, most children have difficulty giving the right amount of money or do not know how to receive the balance from the seller. This is because, when children give a larger amount than the original price, the child will only take the goods purchased after giving money to the seller without the balance. Therefore, children need to have basic knowledge about money such as recognizing and distinguishing between types of money, knowing the value of money, and being able to differentiate the value of a product according to its price.

A child's awareness of the function of money starts at the age of three, but a child still does not have a clear understanding of it [4]. At that age, the person closest to the child is their parent or guardian. Thus, education about money begins with their parents. Then, at the age of seven, children were given pocket money by their parents for school expenses. Therefore, they need to know about money management so that they can be responsible for making decisions for themselves in the future [3]. Besides, the way children appreciate and spend their money is directly influenced by how their parents or guardians teach them. This is because at that time parents were the main source of money for children. All matters of financial decisions in the family should be discussed by parents with their children as they are directly affected by these decisions [5].

The rapid growth of technology is affecting the way people live today. The use of technology in education cannot be denied. Some schools have implemented mobile-based education, such as tablets as one of their teaching mechanisms to make the process more fun and interesting compared to traditional teaching consisting of books and papers. Teaching methods in using mobile games have been successful in motivating students compared to book-based methods as students will be more involved in the process of behavioral, intellectual, and emotional learning [6].

Since it is common nowadays for children to be given devices such as smartphones and tablets by their parents, it is appropriate to use mobile games to educate children on certain topics, especially in recognizing money. It turns out that interactivity in mobile games is effective in learning subjects such as mathematics and science because it is quite difficult to imagine and manipulate concepts through real objects [7]. By using mobile games, the user can avoid any damage to the real money or injuries that may occur during learning sessions if using real money. Moreover, mobile games have interactivity, instant feedback, and risk-free nature that can encourage students to explore and try further in a lesson [8]. Thus, the use of mobile games for education has proven to be beneficial.

3. Background Study

A study has been conducted by [9] to find out the experiences of children learning about money at home and school. While, as agree by [10], young children often have difficulty with understanding money in a traditional formal school learning. Without proper financial education, [11] has claimed that children can be very vulnerable and defenseless; the study agreed that children with poor financial skills often result in delayed financial responsibilities.

The use of mobile games in learning has been widely used nowadays. Most schools are already beginning to implement the use of devices such as tablets in their learning sessions. This is because several advantages can be seen by using mobile games for learning. Among them, the cognitive ability of a child who plays video games at seven years is better compared to children who do not play video games [12].

Also, the use of mobile games in learning has proven successful in motivating students when compared to using the traditional way of learning based on books. At the same time, games like this have a persuasive nature due to their interactive way of use as well as the exciting use of technology [13].

Studies on existing games should be conducted to compare the features of the game and the concept of the game to be used in this research. The concept and pattern of the game are based on learning to know money. Table 1 shows a comparison of each of the existing games that have been developed

IOP Conf. Series: Materials Science and Engineering **979** (2020) 012018 doi:10.1088/1757-899X/979/1/012018

previously, namely Peter's Pig Money Counter, Freefall Money, Renegades Buggies, and MON€¥$! - Money Match Puzzle. Figure 1 shows the existing game interfaces.

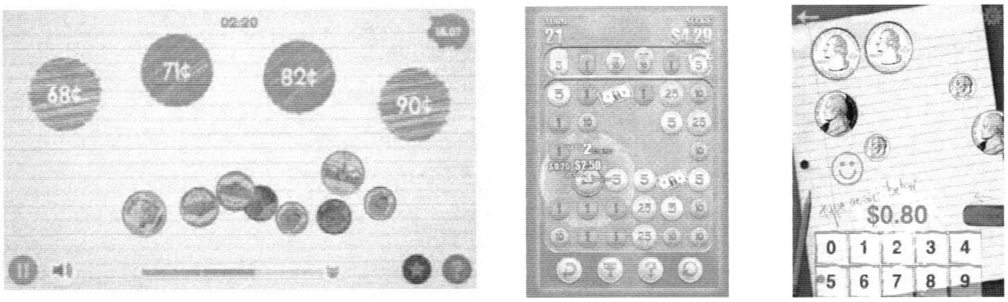

(a) Peter's Pig Money Counter [14] (b) Money Match Puzzle [15] (c) Freefall Money [16]

Figure 1. Existing game to learn about money.

Table 1. Comparison of Existing Game Features.

Application	Features		
	Getting to Know Money	Game Patterns	Interactions
Peter's Pig Money Counter [14]	Some know US coins	Put money in a jar based on the value	Touch, drag, and drop
Freefall Money [16]	Get to know coins using the concept of paper exercises	Write the value of the coin based on the designated coin model	Touch
Renegade Buggies [17]	Learning to collect and save money	Collect money and items you want to buy along the way.	Touch and swipe
MONEY! - Money Match Puzzle [14]	Recognize coins and paper by matching based on design.	Match money with the same shape and value.	Touch and swipe

4. Methodology

To build a mobile game and store children with a basic knowledge of identifying money, the chosen platform is Android. The research has several planned phases, namely Phase 1: Preliminary Research and Data Collection, Phase 2: Design and Implementation, and Phase 3: Testing and Evaluation.

4.1. Phase 1: Preliminary Research and Data Collection

In this phase, the focus of the research is to explore past approaches to money-based learning, including how games are used to achieve interactive learning for children. The activities involve, to find out what syllabus is used and what is the learning methods used in primary school about the basics of money. Data is collected from the textbooks provided.

4.2. Phase 2: Design and Implementation

Based on the information found in the previous phase, the design phase for money base learning games has been implemented to assist children in basic money learning and interactively test their knowledge. The main challenge in this research is to illustrate the syllabus available from books to mobile games. As mentioned above, the target users for this game are students around the age of 7 to 9 years, so mobile games based on touch and finger gestures are appropriate because it makes it easier for children to play this game. The design phase begins when all the requirements have been identified and properly understood. It also involves how the game is viewed by the end-user. From the information obtained from the research phase, it is then used as a design guide for this game. As such, this money-learning mobile game targets consumers seven to nine years. The content for this application is in line with the sub-topics of money for the subject of Primary school syllabus Mathematics. It includes questions about

IOP Conf. Series: Materials Science and Engineering **979** (2020) 012018 doi:10.1088/1757-899X/979/1/012018

knowing money, knowing the value of money, and basic purchase simulations. Thus, the design is divided into sections, as presented in Figure 2.

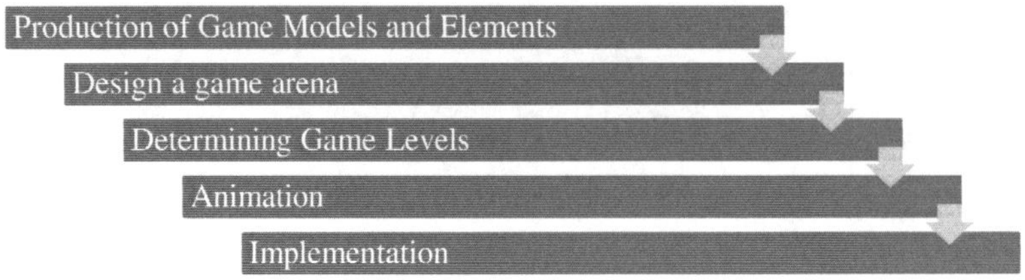

Figure 2. Design process.

After the design phase is complete, a mobile game framework on basic money learning is carried out. A prototype for a 3D puzzle game was developed for this phase. After a complete development phase, this mobile game has been implemented on Android platforms using the Unity IDE.

4.3. Phase3: Evaluation and Testing

This phase discusses the game evaluation of basic money learning for Android platforms. This includes comparisons between existing games and shows improvements if needed. The evaluation process is carried out for the usability and functionality of the game developed. Also, the test phase in this is to identify the complications or errors found in the built game. The game will be tested using a black box testing technique or known as a behavior test and then a user acceptance test. Figure 3 shows the Research Methodology.

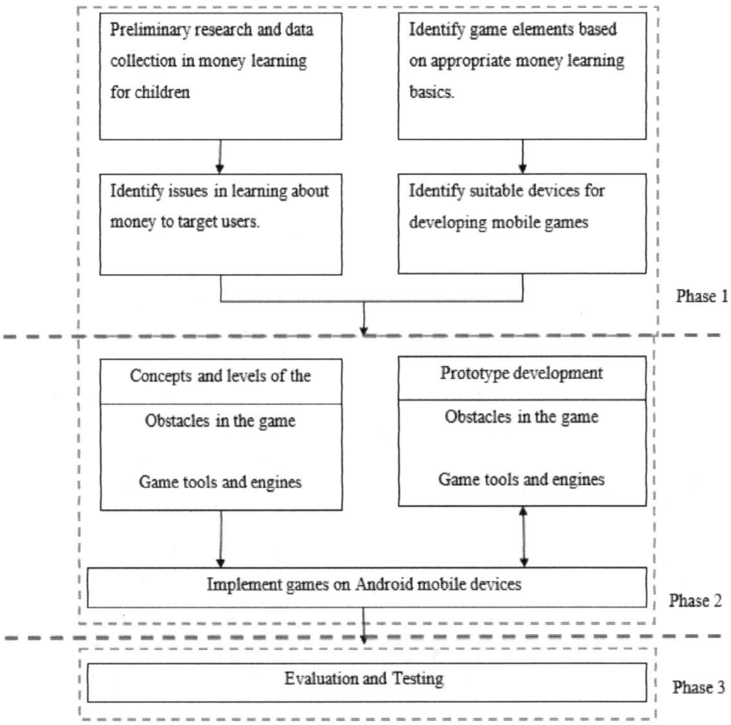

Figure 3. Research Methodology.

5. Game Design

The concept of this game is a 2-Dimensional puzzle; questions about the basics of money were asked to the player. The game has three different types of games, namely recognizing money by matching the front and back of the money, calculating the value of money by choosing the correct answer, and also the purchase of items by paying the right amount. Figure 4 shows the framework of this game.

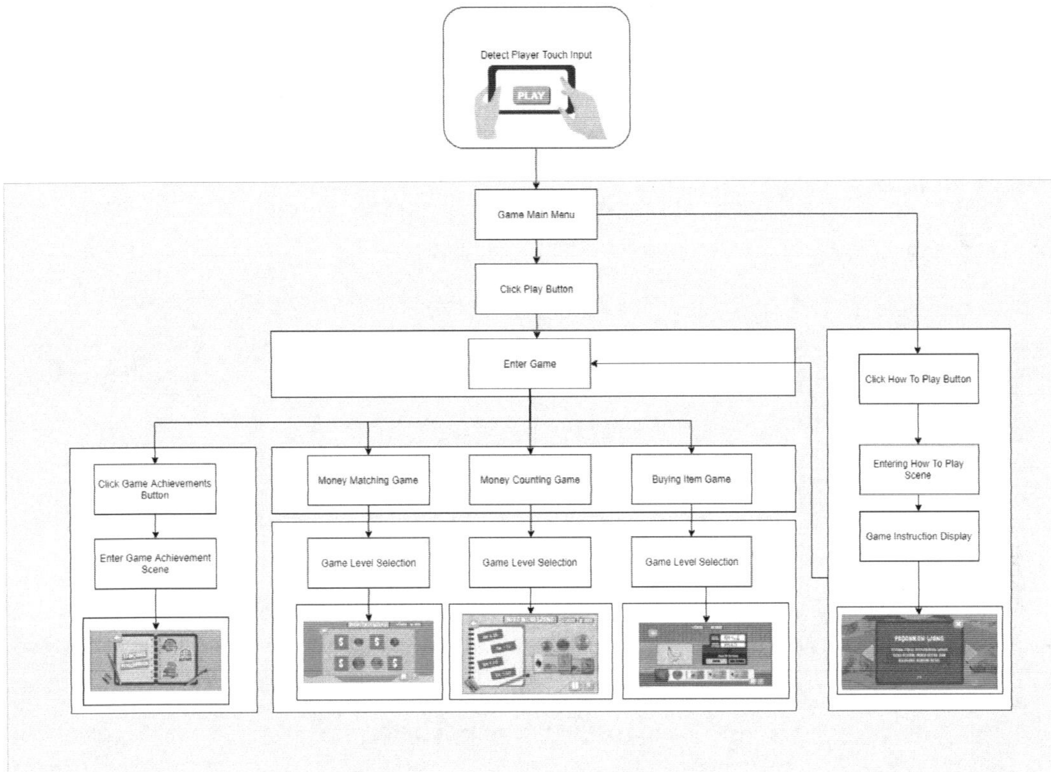

Figure 4. Game Framework.

Before playing, players can also enter the screen of how to play where it will explain about this game. When players press the start button, they will enter a game type selection menu where they can choose between the three desired game types. In this menu as well, they can go to the awards list view to see the awards won if they complete each type of game by completing each level. Next, the first game is to match the money according to the same design. There are two designs for coins, the front design, and the back design. Next, players also need to match the front design of the banknote and the back design of the banknote. The last stage is a combination of these two designs. Each level has its time limit.

The next game is to determine the value of the money displayed. A combination of banknotes and coins will be displayed, and players are required to calculate the value of the money. Choices of answers will be given to the player.

Finally, a game was developed for the purchasing items. Players will be given items at a certain price, and they will have to pay according to the total price of the item stated. A list of money will be given to the players so that they can add up each money to make the desired amount.

5.1. User Interface Example

There are three buttons, namely the Start button to start the game, the How to Play button where it will give a little description about this game, and the Exit button to exit the game. Figure 5 shows the main interface of the first view once the player enters the BijakWang game.

IOP Conf. Series: Materials Science and Engineering **979** (2020) 012018 doi:10.1088/1757-899X/979/1/012018

Figure 5. Main Menu Interface.

6. Implementation

There are three mini-games, namely matching money, counting money, and purchasing the items. Each game has a different way of programming and game elements.

6.1. Matching Game

Players need to match the money front design to the right money back design. Correct answers will be given marks. There is also a time count for this game. Figure 6 shows a game for matching money.

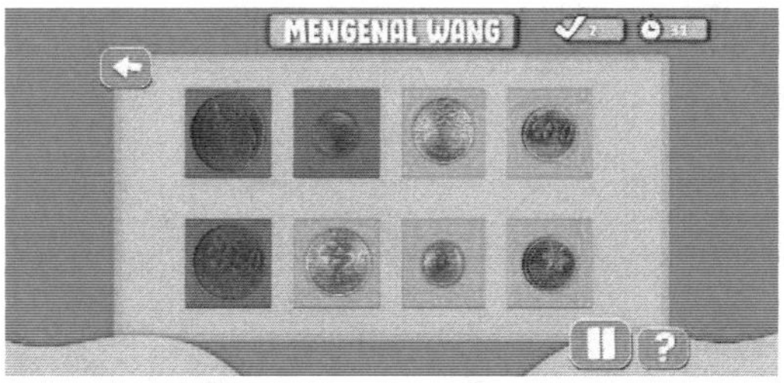

(a) Matching the coins

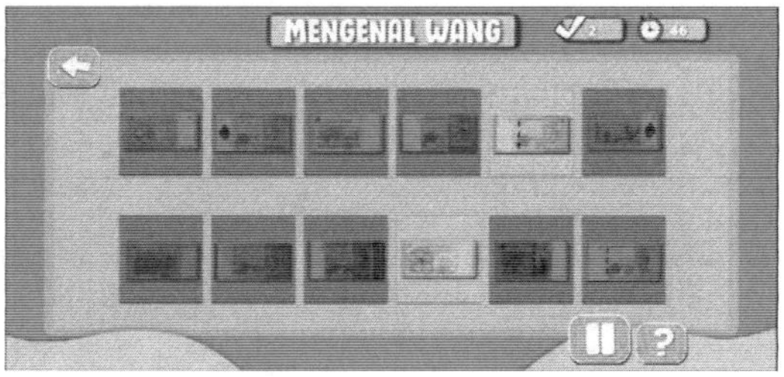

(b) Matching the money notes

Figure 6. Matching Money.

6.2. Counting Money

The game of calculating the value of money is for the player to choose the value of the correct answer based on the pieces of money given. Figure 7 shows the game to calculate the total value of money.

(a) The main interface for counting money

(b) If the user picks the correct amount

Figure 7. Main Menu Counting Money.

6.3. Purchasing Items

In this game, the player must pay the correct amount of money based on the price of goods spawned at random. The method of payment is by tapping at any of the money shown on the screen and the value in the Total Payment Amount will increase based on the value of money. Players will add up the value of money until they get a sufficient amount for payment. When a player presses the Pay button, the player will know whether the answer is correct or incorrect. Figure 8 shows the purchasing items or goods gameplay where the player can spend money to buy the item shown. In Figure 8 (a), a random item was spawned on the left-hand side of the screen while the right-hand side shows the cash register machine with the price of the item, and at the bottom of the screen are the choices of money for the player to tap. Figure 8 (b) shows the correct answer feedback once the player answers the right amount of money. The player will receive points and proceed to the next question. However, Figure 8 (c) shows the incorrect answer feedback once the player answers an incorrect amount of money. The player will not gain any point and will answer the question again until it is correct.

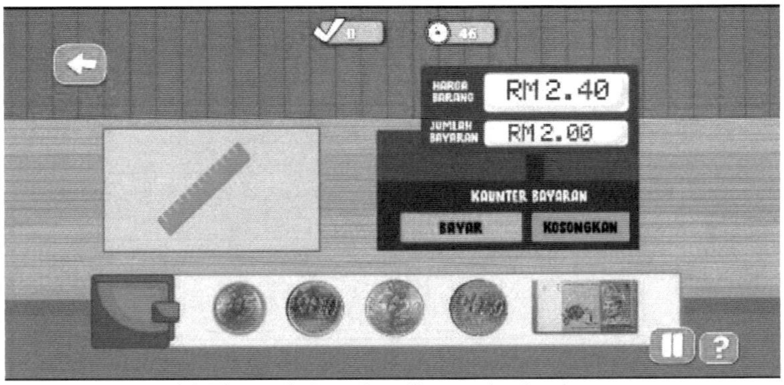

(a) The main interface for purchasing items

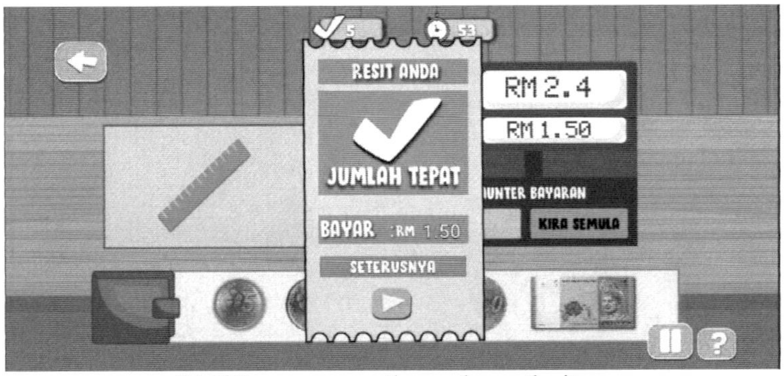

(b) If the user correctly purchases the items

(c) If the user provides an incorrect amount to purchases the items

Figure 8. Purchasing the items.

7. Evaluation

The evaluation has been performed to observe usability testing. User satisfaction while using this research is also assessed to see whether the game is accepted by the user [18]. Target users will be asked to play the game and complete the same requested task to see if they will have any problems. Figure 9 shows the ongoing test to one of the target users of this game, namely children between the ages of seven (7) to (9) years.

IOP Conf. Series: Materials Science and Engineering **979** (2020) 012018 doi:10.1088/1757-899X/979/1/012018

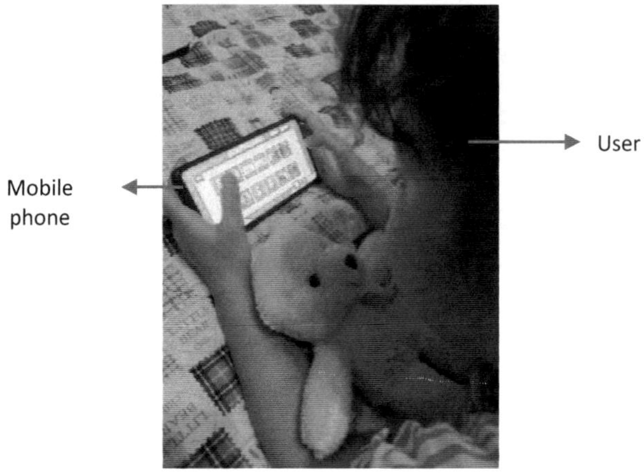

Figure 9. Mobile games with the user.

This test is conducted to see if this research can meet the needs of the target users. Eight respondents have filled out the form with the help of their parents. The first question is, the children were asked to answer Q1: I can learn to identify money with this game and the result as in Figure 10. Six respondents have successfully managed to match the money correctly.

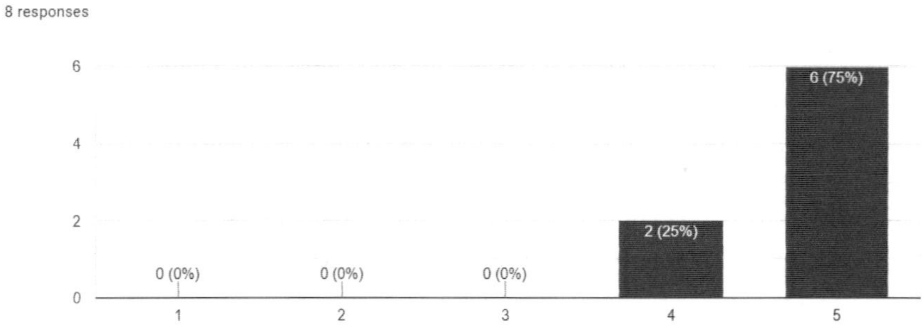

Figure 10. The user responds to the matching game.

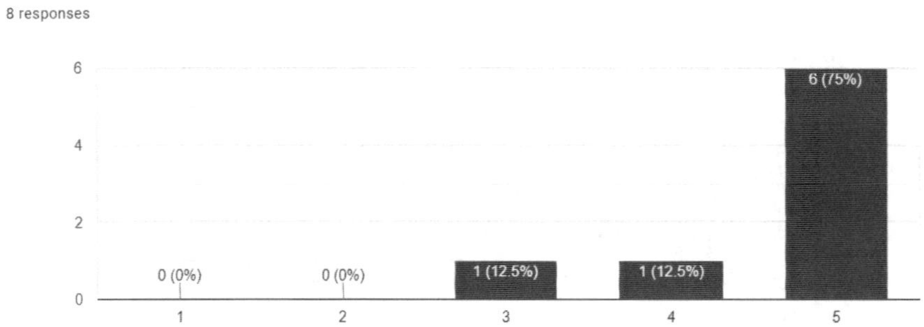

Figure 11. The user responds to the game for counting money.

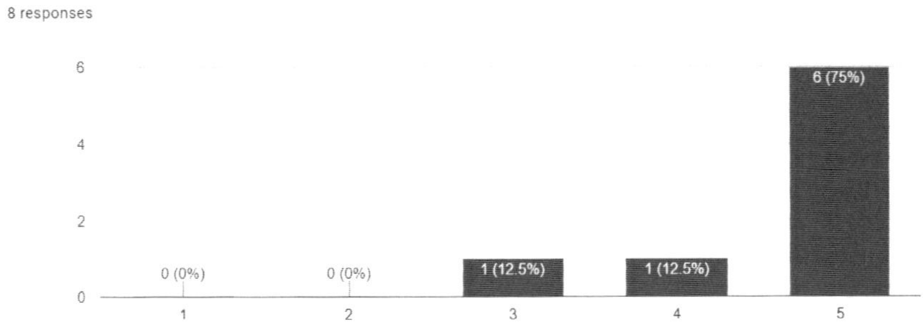

Figure 12. The user responds to the game for purchasing the item.

The second question is, the children were asked to answer Q2: I can learn to count the money with this game and the result as in Figure 11. Six respondents have successfully learned to count the money adequately. As in Figure 12, the result for Q3. The third question the children were asked to answer Q3: I can use the money to purchase the items. Six respondents have used the money to purchase the items and found themselves able to purchase the item with the exact amount of money.

8. Conclusion

A research is complete when it has achieved the goals and objectives of the research. The goal of developing mobile games to get to know money, knowing the types of money, and basic purchasing skills for children aged seven to nine years has been achieved. Apart from that, the three stated objectives can also be achieved. They are starting with the first objective, which is to analyze the needs in the development of mobile games for learning to know money for children. Next, to develop mobile games for children aged seven to nine years that can help them learn the basic knowledge of money is also successful.

This research will only be implemented on Android platforms. This is because smartphones that use the Android platform are accessible to get by users. Also, this research is only for single players because it focuses on education. Next, lacking in-game animation to engage players more effectively. Finally, this research has only three stages which include easy, medium, and difficult due to time constraints.

For future work, there are some plans to improve this Smart Money mobile game research. The first is to add more animation so that it can attract players more effectively. Also, to add elements of 3-Dimensional (3D) objects, especially in displaying the purchase of items because the application only in 2D. The rewards can be revealed, besides the games can be converted into Augmented Reality (AR) application.

Acknowledgments

We would like to express our appreciation to ViCubeLab at Universiti Teknologi Malaysia (UTM) for the technical supports and facilities. We are also thankful to the parties that contributed the fundamental knowledge in this paper.

Reference

[1] Portia, O. 2011. *Barter trading: An empirical investigation of management practices. African Journal of Business Management*, 5(31).
[2] Tang, T. L. P. 2010. *Money, the meaning of money, management, spirituality, and religion. Journal of Management, Spirituality & Religion*, 7(2), p 173-189.
[3] Wahab, R., Sabri, M. F., and Ramli, N. N. 2016. *Money Literacy, Shopping and Purchasing Patterns among Primary School Students in Malaysia. Journal of Education and Human Development*, 5(1).
[4] Te'eni-Harari, T. 2016. *Financial literacy among children: the role of involvement in saving money. Young Consumers*, 17(2), p 197-208.

[5] Dahari, Z. B., and Khurshid, A. 2013. *Money Awareness among Malay Children (EDUKIOMONEY): An Exploratory Study in Malaysia. International Journal of Business and Social Science*, 4(6).
[6] Elaish, M., Abdul Ghani, N., Shuib, L., and Al-Haiqi, A. 2019. *Development of a Mobile Game Application to Boost Students' Motivation in Learning English Vocabulary (Vol. PP).*
[7] Mitchell, A., and Savill-Smith, C. 2004. *The use of computer and video games for learning: A review of the literature.*
[8] Kirriemuir, J. 2002. *The relevance of video games and gaming consoles to the Higher and Further Education learning experience.*
[9] Brenner, M. E. (1998). Meaning and money. Educational Studies in Mathematics, 36(2), 123-155.
[10] Fanelli, B. H., & Tracy, D. M. (1995). Learning about Money: Effects of a Cognitively Appropriate Curriculum on Second-Grade Students and Teachers.
[11] Martin, A., & Oliva, J. C. (2001). *Teaching children about money: Applications of social learning and cognitive learning development theories*. Journal of Family and Consumer Sciences, 93(2), 26.
[12] Arias, M. 2014. *Using video games in education. Journal of Mason Graduate Research*, 1(2), p 49-69.
[13] Elaish, M., Abdul Ghani, N., Shuib, L., and Al-Haiqi, A. 2019. *Development of a Mobile Game Application to Boost Students' Motivation in Learning English Vocabulary (Vol. PP).*
[14] Peter Pigs Money Counter. (2020). Retrieved from http://practicalmoneyskills.com/play/peter_pigs_money_counter
[15] Money Match Puzzle. (2014). Retrieved from https://itunes.apple.com/en/app/mon-$!-money-match-puzzle/id906709425
[16] Freefall Money. (2020). Retrieved from https://www.commonsense.org/education/app/freefall-money
[17] Renegade Buggies. (2020). Retrieved from https://www.commonsensemedia.org/app-reviews/renegade-buggies
[18] Mah, K. 2017. *Evaluating the Usability and Ease of Use of a Mobile Game to Enrich Understanding of Hawaiian History and Culture for 4th Grade Students.*

IOP Conf. Series: Materials Science and Engineering **979** (2020) 012019 doi:10.1088/1757-899X/979/1/012019

Math teaching aids in shape topics for PT3 students

R E Roshidan[1]*, N A F Ismail[1], N Z S Othman[1] and N M Suaib[1,2]

[1] School of Computing, Faculty of Engineering, Universiti Teknologi Malaysia, 81310 Skudai, Johor Bahru, Malaysia
[2]UTM Big Data Center, Ibnu Sina Institute of Scientific and Industrial Research, Universiti Teknologi Malaysia, 81310 UTM Skudai, Johor, Malaysia.

* rashad006@gmail.com

Abstract. Mathematics is a significant pillar in science as mathematics has built technology, machines and digital as progress in this century. Therefore, mathematics is essential to practice since childhood. Although various efforts have been made in updating the teaching methods to students by teachers, students still have difficulty in understanding mathematical concepts, especially in the topic of form. Most students memorize more than they know logic and discoveries in mathematics. Teaching aids are interactive software that can be used simulated. Therefore, the goal of this project is to develop mathematics teaching aids for shape topics for Form Three Assessment (*Pentaksiran Tingkatan Tiga* - PT3) students. The study involves three main phases which are initial research and data collection, design and implementation and assessment.

1. Introduction

Computer technology is developing as development in the field of computers is emphasized because computers are now widely used by people around the world as a routine and daily life. Along with that, software technology is also being renewed and developed especially for the young.

In learning, teachers face challenges in teaching where each individual has a different intellectual understanding from each other [1]. This challenge should be addressed by using technology, such as the use of teaching aids for subjects that teachers handle. According to Tolbert [2], the use of teaching aids shows a positive effect on student achievement, where students can easily understand the concept of learning.

Furthermore, instructors also use teaching aids along with computers in helping weak students for certain subjects, especially mathematics. This is because there are students who have problems in understanding the basics of mathematics from the beginning despite knowing the calculation for the question. [3] has investigated the metacognitive behaviour of students in solving mathematical problems of Form Three Assessment (*Pentaksiran Tingkatan Tiga* - PT3). In Malaysia, the students in the age range of thirteen to fifteen years will sit the PT3 examination. A recent study in [4] claimed that the Higher-order thinking skills (HOTS) mathematical problem-solving was first introduced in the PT3 in 2014. However, so far, there have been no studies of students' ability to solve the mathematical problems in PT3 assessment. Therefore, this paper aims to describe the development of teaching aids in three-dimensional (3D) for shape topics in mathematics, and a case study is the PT3 students.

2. Problem Background

Nowadays, technology, machines and digital have introduced new knowledge and discoveries. All of this starts from mathematics since it was first discovered and studied by some of the earlier figures for so long. To date, mathematics has changed our environment directly or indirectly in our daily lives. Therefore, mathematics is very important to be taught since kindergarten.

Teachers today strive to improve their teaching methods so that the learning objectives for students are more successful. The main goal of teachers in teaching mathematics lessons is to help students' proficiency in mathematics, that is, the ability to understand, evaluate, and apply mathematical knowledge regardless of any situation related to mathematics. Based on PT3, the constructs that need to be given to students are such as skills to remember, understand, apply, solve problems and evaluate through written tests and types of items such as Multiple-Choice, *Objektif Aneka Pilihan* (OAP), *Objektif Pelbagai Bentuk* (OPB) and *Subjetif Respons Terhad* [5]. Mathematical topics such as shapes are very relevant and require those skills.

However, students are more likely to memorize formulas to perform calculations in answering mathematical questions than to understand the intent or concept of the question. Although students are proficient in making calculations, students will have difficulty in choosing the formula that needs to be used to solve mathematical problems. This is because students memorize the facts or how to use the formula alone without feeling the mathematics as a process of discovery and exploration in depth. For students, mathematics is known as a call-and-response game [6]. Furthemore, for students, it is difficult to imagine the proper view of the shape by looking at static image in text book.

Therefore, there are several other methods used to teach students in the field of mathematics in line with the era of technology. Among them is interactive software where mathematics is taught in a simulated calculation controlled by the player. The simulation software provides space for students to explore more deeply about mathematics while stimulating interest. Examples of such software can be found on the Math is Fun website, where the site provides several types of mathematical software from various subjects, such as Broken Calculator, where players have to answer number operations based on the function of the calculator button provided only [7]. This software provides challenges and space in exploring each solution that can unravel the given questions.

In conclusion, the problem of the frequency of memorizing the use of formulas without the internal exploration of the subject will make it difficult for students in understanding the subject of mathematics such as shape. Among other teaching methods to curb this problem is to use interactive software taught in numerical simulation to increase students' interest in exploring solutions to each mathematical problem. Thus, teaching aids can be used to assist students in mathematics subjects for shape topics.

3. Background Study

Studies on existing software are conducted to compare the characteristics of teaching aids and the concept of teaching aids that has been used in this study. The pattern of teaching aids software is important to identify the concept of teaching aids that are appropriate to the target user, which are PT3 students in Table 1.

Table 1. Mathematic teaching aids software comparison for shape topic.

Software	Features		
	Operation	Software Patterns	Special Features
Open Source Physics	The volume formula for a rectangular pyramid	Change the shape of the pyramid based on the value of area and height	Allows the user to observe the shape of the pyramid as a whole
Geogebra	The volume formula for cylinders	Change the shape of the cylinder based on radius and height	Has a detailed operating indication of the volume
Math Interactives	No Operations	Quiz or simulation to show shape by surface area	Produced in 3D and this type of software is in the form of simulations and quizzes.

4. Project Methodology

In order to build mathematics teaching aids for PT3 students, this project use computer-based, and involves several phases methodology, as shown in Figure 1.

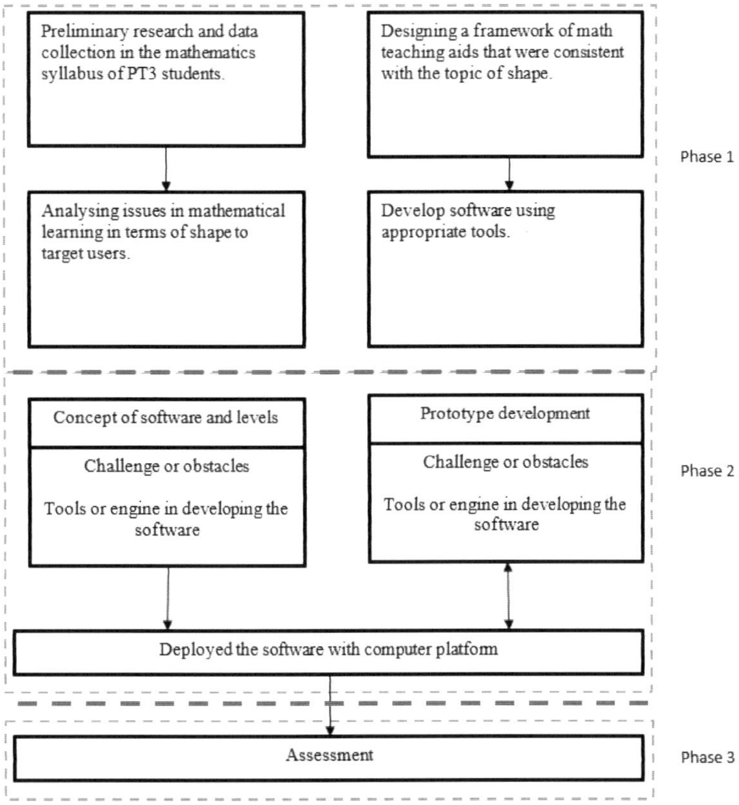

Figure 1. Project Methodology.

Phase 1 is focused on exploring past approaches to develop teaching aids for mathematic in the topic of shape, including how their software achieve interactive learning for PT3 students in determining basic formulas and operations. PT3 syllabus in mathematics has been collected, data were collected by referring to books to find out how to learn secondary school and basic arithmetic operations.

Based on the information found in the previous phase, phase 2 is the invention phase for mathematic teaching aids software in terms of shapes is needed to assist in the learning of shape topics interactively. The main challenge in this project is to illustrate the syllabus available from books to 3D software.

Phase 3 discusses the evaluation of mathematic teaching aids in terms of shape, Shape Lab for computer platforms. This includes comparisons between existing teaching aids software and showing improvements if needed. The evaluation process is carried out for the usability and functionality of the software developed.

4.1. Description of Technology Used

The technologies used to build this software are Unity and Blender. Unity is a real-time engine developed by Unity Technologies in June 2005. The Unity engine is capable of producing applications, systems or software either 2D or 3D. This technology is used to produce 3D computer software using C ++ or C # programs to produce the required functions according to the scope of this project.

Blender is a free and open-source 3D computer graphics software. The use of Blender is to create animations, visual effects, they create models and so on. Blender software can be used to facilitate the software development process of math-specific teaching aids in Unity engines. Most of the forms used for this project are produced by the Blender program. Table 2 list the hardware requirements for this project and Table 3 list the software requirements for this project in general.

Table 2. Hardware Requirements.

Hardware	Model
Computer/Laptop	ASUS TUF GAMING FX505
Central Processing Unit(CPU)	Intel Core i5 8th Generation
Graphic Processing Unit (GPU)	Nvidia GeForce GTX 1050
Random Access Memory (RAM)	12 GB
Hard Drive Storage	1 TB

Table 3. Software Requirements.

Software	Model/Version
Operating Systems	Windows 10
Application Development	Unity 2019.1, Visual Studio 2017
3D Modelling Engine	Blender

5. Design

The concept used for this project is to learn the concept where the user puts the answer key in the keyboard and sees a simulation of how the shape is designed, involving multiplication for volume calculation.

In the early stages, users need to press the play button on the page to use Shape Lab software. Users will see tutorials on how to use the software such as how to run, and answer questions. After that, they will start playing on the first level which is the square level. The main instructions for this software are about movement and how to answer the design for squares, triangles and circles.

There are six shapes for each type of shape displayed in a wireframe line, to indicate the shape is a question that needs to be filled. Each shape has variations in terms of scale on each part of the formula such as length, height and width. Therefore, each shape does not have the same length, height and width values. The framework of teaching aids software, Shape Lab is shown in Figure 2.

5.1. Home Page Interface

The interface design, or referred to as the interface design, refers to the information displayed to the player as a guide on how to play Shape Lab. When the user opens this software, the interface provides two menus, namely Play and Quit. Users need to press the Play button to start using this software in Figure 3.

5.2. Shape Question Interface

The user needs to enter all the answer values , i.e. length, height and width of the rectangle, as shown in Figure 4 (a). When the answer value is given, the shape will be designed in the preview provided in the middle of the question page. Each value of the answer will affect the appearance of the answer to that shape. When the user fills in all the answer values, the player has to press the Apply button to check the answer. If all the answers are correct, the question is closed, and the shape for the question is displayed in blue, as shown in Figure 4 (b).

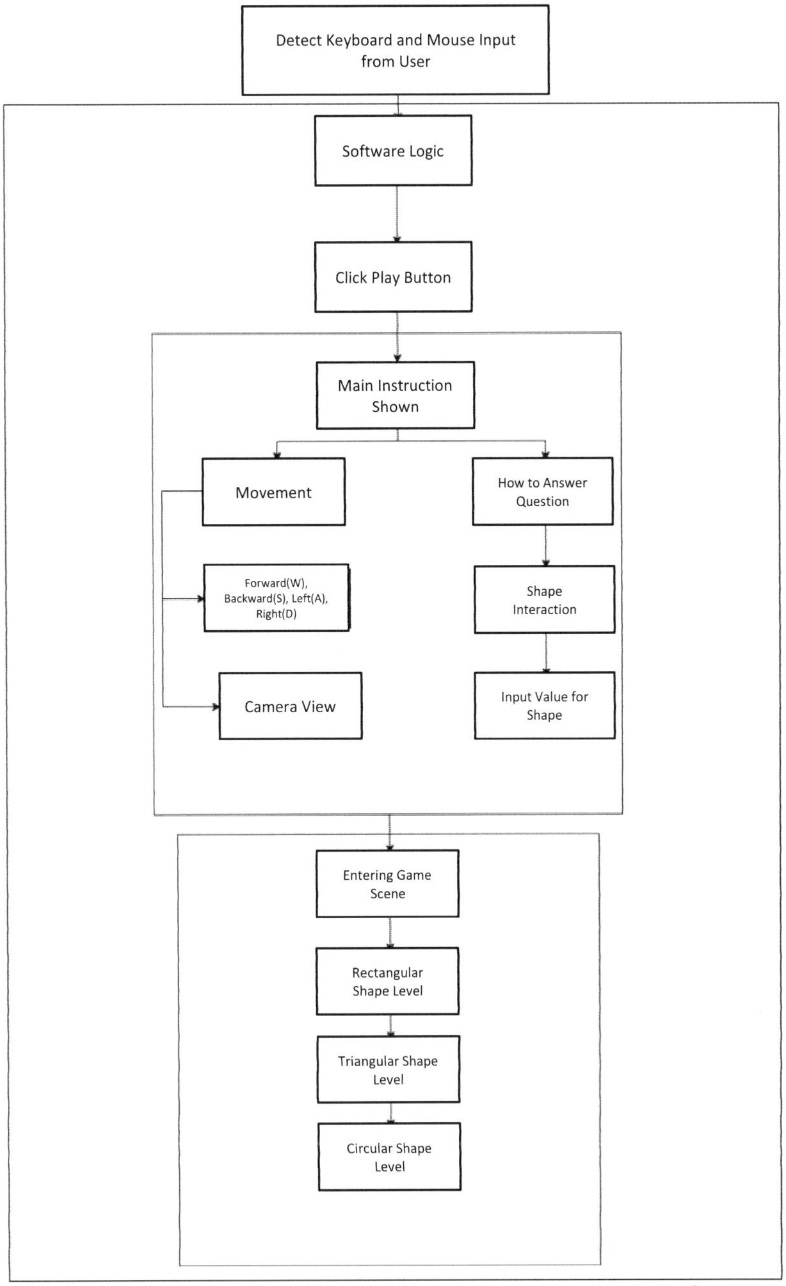

Figure 2. Shape Lab framework.

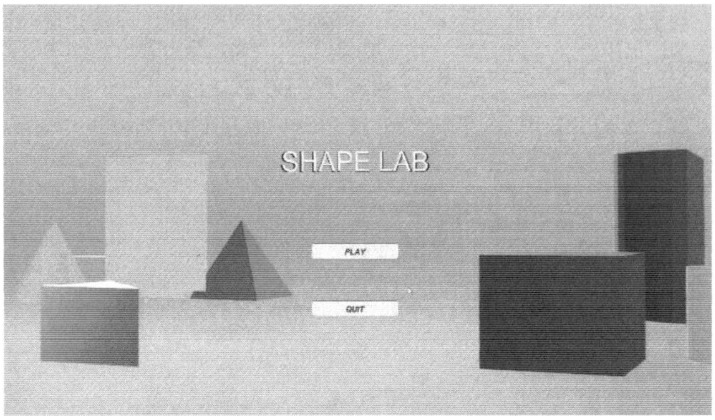

Figure 3. Shape Lab software homepage.

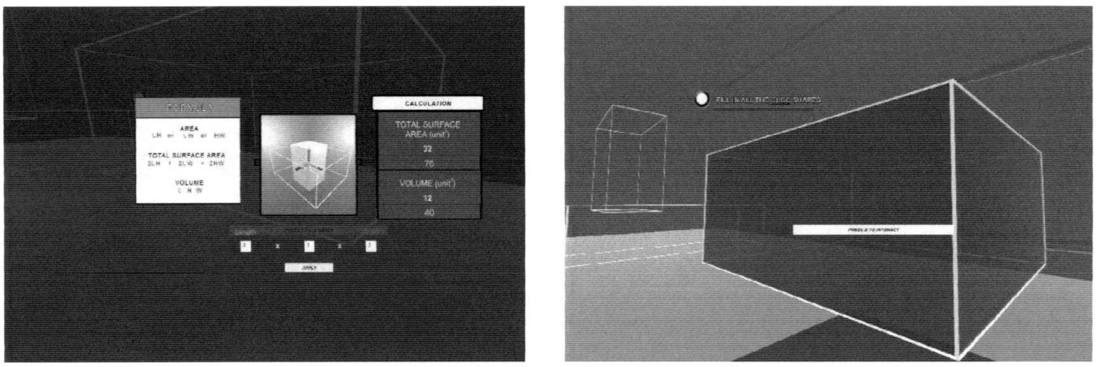

(a) Question interface (b) Successfully answered questions

Figure 4. The Application.

6. Implementation

There are three main implementations in developing this software, namely character movement and camera view, and arithmetic shape calculation interactions along with simulations and calculations for calculated displays for area and volume.

6.1. Implementation of Character Movement and Camera View

Basically, in teaching aids software should involve character movement and camera views to provide a 3D space experience to the user. In addition, character movement can be controlled using the keyboard provided by Unity Engine [15]. Generally, to move an object, we need to change the transformation value in terms of translation such as X and Z. To change the translation value of X from the user character, the float variable x is used and takes the input of the Horizontal keyboard, namely W and S. To change the value translation Z, float z variables are used and take the input of Vertical type keyboards, namely A and D. All of these coding implementation is shown in Figure 5.

```
//Gravity Ground Check
isGrounded = Physics.CheckSphere(groundCheck.position, groundDistance, groundMask);


if(isGrounded && velocity.y < 0)
{
    velocity.y = -2f;
}



float x = Input.GetAxis("Horizontal");
float z = Input.GetAxis("Vertical");

Vector3 move = transform.right * x + transform.forward * z;

controller.Move(move * speed *Time.deltaTime);

velocity.y += gravity * Time.deltaTime;

controller.Move(velocity*Time.deltaTime);
```

```
float mouseX = Input.GetAxis("Mouse X")*mouseSensitivity*Time.deltaTime;
float mouseY = Input.GetAxis("Mouse Y")*mouseSensitivity*Time.deltaTime;

xRotation -= mouseY;
xRotation = Mathf.Clamp(xRotation, -90f, 90f);

transform.localRotation = Quaternion.Euler(xRotation, 0f, 0f);
playerBody.Rotate(Vector3.up * mouseX);
```

Figure 5. Coding for movement and camera rotation.

6.2. Implementation of Arithmetic Shape Calculation Interaction with Simulation

Figure 6 shows the variables and functions that are closely related to user input in answering shape questions. The variables in the *CubeVolume.cs* script are divided into length, width and height. Then, based on the answer input box, he will call the function together with the respective variables.

For example, when the user changes the width value, the width function, i.e. void *widthChanged* (string textw), will be called and change the appearance of the width for the following simulation shape. If there is no input, the shape of each type of value will be given to the default value (default value) which is 0.01.

```
public void widthChanged(string textw)
{
    if (textw == "0")
    {
        questionOject.transform.localScale = new Vector3(tempLength, tempHeight, tempWidth);
    }
    else
    {
        float width = float.Parse(textw);
        tempWidth = width;
        questionOject.transform.localScale = new Vector3(tempLength, tempHeight, width);
    }
}
```

```
public void widthChanged(string textw)
{
    if (textw == "0")
    {
        questionOject.transform.localScale = new Vector3(tempLength, tempHeight, tempWidth);
    }
    else
    {
        float width = float.Parse(textw);
        tempWidth = width;
        questionOject.transform.localScale = new Vector3(tempLength, tempHeight, width);
    }
}
```

```
public void widthChanged(string textw)
{
    if (textw == "0")
    {
        questionOject.transform.localScale = new Vector3(tempLength, tempHeight, tempWidth);
    }
    else
    {
        float width = float.Parse(textw);
        tempWidth = width;
        questionOject.transform.localScale = new Vector3(tempLength, tempHeight, width);
    }
}
```

Figure 6. Coding Simulation Calculations for Rectangular Shapes.

6.3. Implementation of Calculation for Calculation Display for Area and Volume.

Based on Figure 7, when the user places the answer value in one of the answer fields provided, the *calculation()* function is called to identify whether the answer value is a long, high or wide value. If only one answer value is given from the three answer fields, the calculation of area and volume is not taken into account. However, if only two answer values are given from any of the filled answer fields, the surface area will be calculated. If all the answer values are entered, the overall calculation of the surface area and volume of the shape is used. At the end of the function, the answer to the calculation made will be displayed on the calculation interface.

```
void calculation(float length, float height, float width)
{
    if (length < 1 && height >= 1 && width >= 1)
    {
        area = height * width;
        volume = 0 * height * width;
        surfaceAreaUI.SetActive(false);
    }
    else if (height < 1 && length >= 1 && width >= 1)
    {
        area = length * width;
        volume = length * 0 * width;
        surfaceAreaUI.SetActive(false);
    }
    else if (width < 1 && length >= 1 && height >= 1)
    {
        area = length * height;
        volume = length * height * 0;
        surfaceAreaUI.SetActive(false);
    }
    else if (length >= 1 && height >= 1 && width >= 1)
    {
        float surfaceAreaValue = ((height * length) * 2) + ((width * length) * 2) + ((height * width) * 2);
        area = surfaceAreaValue;
        volume = length * height * width;
        surfaceAreaUI.SetActive(true);
    }
    else if (length < 1 && height < 1 && width < 1)
    {
        area = 0f;
        volume = 0f;
    }
    else if ((length < 1 && height <1 && width >= 1) || (height < 1 && width < 1 && length >= 1) || (width < 1 && length < 1 && height >= 1))
    {
        area = 0f;
        volume = 0f;
    }
    else
        wrongQuestion();

    areaValueText.text = area.ToString();
    volumeValueText.text = volume.ToString();

}
```

Figure 7. Coding for area calculations and volumes for rectangular shapes.

7. Assessment Results

This study has gathered the target users for math teaching aids software in terms of shape, and the study was conducted at ABIM Primary and Secondary School (SERI) and Secondary School (SEMI) in Kampung Sungai Ramal Dalam. However, due to time constraints and the coronavirus impact, the government has announced the Movement Control Order (PKP), the study has been carried out to gain access to only one school. The results have been formed in graph, and the 15 respondents were asked to answer the questionnaire.

Figure 8 shows the eight respondents have satisfied with the software in term of easy to use with the percentage of 53.3%. However, four respondents gave a slightly satisfied rating with 26.7% and 3 respondents gave a neutral rating with a percentage of 20%.

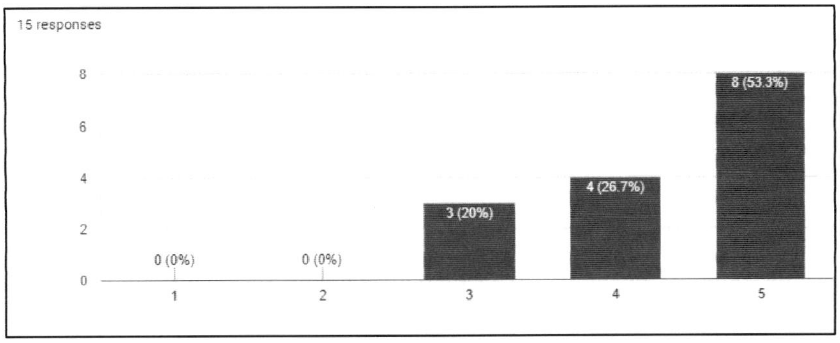

Figure 8. The user responds for easy to learn when using Shape Lab software.

Figure 9 below shows the simulation is suitable for learning. Ten respondents gave a satisfied rating with a percentage of 66.7%. Meanwhile, four respondents gave a slightly satisfied rating while three respondents give a neutral rating, with a percentage of 26.7% and 20% respectively.

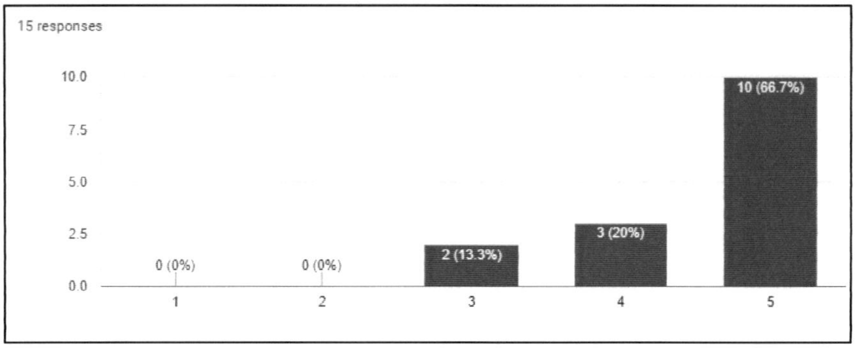

Figure 9. Suitability of simulations used in Shape Lab software.

Finally, Figure 10 shows nine respondents gave satisfied rating with the software in term of ease of use with the percentage of 60%. Three respondents which are 20% of all of the respondents gave a slightly satisfied rating, and two respondents gave neutral rating with a percentage of 13.3%.

The evaluation of Shape Lab software showed positive feedback. Most users choose the strongly agree option for each question indicating that most users agree with the possibility of learning quickly when using the math teaching aid software in terms of shape, Shape Lab. Also, most users find that the simulations provided in the Shape Lab software are appropriate and can give the user a quick exposure of the shape.

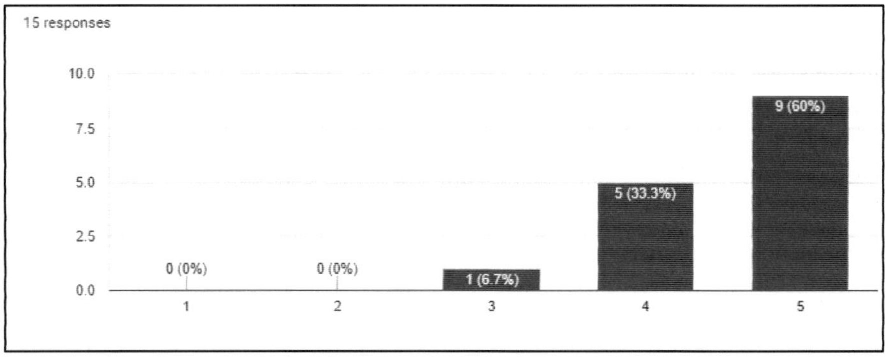

Figure 10. Ease of use when the user works with Shape Lab software.

This software will be tested using black-box testing, and user evaluation to test the effectiveness of mathematics teaching aids software on the target users, PT3 students.

The results for the black-box testing as presented in Table 4, to check the user feedback on the functionality of the application. The verification of the value of the given answers and the calculation of calculations for each type of shape.

Table 4. Black-box testing results in Shape Lab software.

Item	Description	Expected Results	Actual Results
1	The Play button is pressed to start the software.	Page scenes are converted to game scenes	✓
2	The WASD key is pressed to move	Based on the pressed keyboard keys, it will move forward (W), back (S), left (A) and right (D)	✓
3	Move the mouse when in a game scene	Mouse movement allows the player character to look around	✓
4	Approach the question mark symbol	Game guides will be displayed if nearby.	The first symbol indicates the level ✓ The second symbol shows how to answer the question ✓
5	Approaches shape questions based on the level of player being.	A question interaction message will be displayed to indicate the player can answer the shape question when close enough.	All shape questions in Shape Lab software ✓
6	Press the E key when close enough to the question.	Displays questions based on shapes close to the player.	✓ (This interaction can only be made if there is no input from another key.)

8. Conclusion

The aims of this project to developed the application to learned mathematics in terms of shape for secondary school students PT3, *Pentaksiran Tingkatan Tiga* (PT3) 16], successfully achieved. The learning and information is transformed into attractive and interactive using graphical visualization, which is help the students learning journey most easier and enhance the students understanding.

Acknowledgement

This work was supported by the UTM Encouragement Research grant, Universiti Teknologi Malaysia, Malaysia (Q.J130000.2651.17J61). Special thanks to ViCubeLab at Universiti Teknologi Malaysia (UTM) for the technical supports and facilities. We are also thankful to the parties that contributed the fundamental knowledge in this paper.

Reference

[1] Hamdan, A. R., and Yasin, H. (2010). *Penggunaan Alat Bantu Mengajar (ABM) Di Kalangan Guru-Guru Teknikal Di Sekolah Menengah Teknik Daerah Johor Bahru, Johor.*

[2] Tolbert, E. (2015). *The Impact of Computer-Aided Instruction on Student Achievement.*

[3] Abdullah, A. H., Ahmed, S. A., Rahman, S. N. S. A., Mun, S. H., & Mokhtar, M. (2017). Metacognitive Behaviour of Malaysian Students While Solving Mathematical Problems of Form Three Assessment (PT3). Bolema: Boletim de Educação Matemática, 31(59), 907-927.

[4] Abdullah, A. H., Fadil, S. S., Abd Rahman, S. N. S., & Tahir, L. M. (2019). Emerging patterns and problems of higher-order thinking skills (HOTS) mathematical problem-solving in the Form-three Assessment (PT3). South African Journal of Education, 39(2).

[5] *Format Pentaksiran Tingkatan Tiga (PT3) Mulai 2019.* 2018. Retrieved 22 May 2019, from http://lp.moe.gov.my/index.php/peperiksaan-pentaksiran/peringkat-menengah-

rendah/pentaksiran-tingkatan-3-pt3/744-format-pentaksiran-tingkatan-3-pt3-mulai-2019
[6] Orlin, B. 2013. *When Memorization Gets in the Way of Learning.* from https://www.theatlantic.com/education/archive/2013/09/when-memorization-gets-in-the-way-of-learning/279425/
[7] Pierce, R. 2019. *Math is Fun.* from http://www.mathsisfun.com/
[8] Alan. 2019. *Shapes! A Geometry Activity for Children | ABCya!* Retrieved 1 May 2019, from https://www.abcya.com/games/shapes_geometry_game
[9] Hoe, I., Jin, J. H., Mun, Y. C., Zahir, N. A. M., and Muhammad, N. N. F. N. 2019. *KSSM &PT3 MATEMATIK Tingkatan 1, 2 & 3*: PNI Neuron (M) Sdn. Bhd.
[10] *Math vs Monsters - Play with Math Games.* (2019). Retrieved 1 May 2019, 2019, from https://www.mathgames.com/play/mathvsmonsters.html
[11] Orlin, B. 2013. *When Memorization Gets in the Way of Learning.* from https://www.theatlantic.com/education/archive/2013/09/when-memorization-gets-in-the-way-of-learning/279425/
[12] Purohit, K. C. 2016. *Mathematics Science Of Pattern Shapes And Number.*
[13] Riccitiello, J. 2014. *John Riccitiello sets out to identify the engine of growth for Unity Technologies In D. Takahasi (Ed.)*, GamesBeat 2014 conference. VentureBeat.
[14] *Format Pentaksiran Tingkatan Tiga (PT3) Mulai 2019.* 2018. Retrieved 22 May 2019, from http://lp.moe.gov.my/index.php/peperiksaan-pentaksiran/peringkat-menengah-rendah/pentaksiran-tingkatan-3-pt3/744-format-pentaksiran-tingkatan-3-pt3-mulai-2019
[15] Jafree, S. N. S., Hussin, N. H. A. C., and Azhari, S. A. 2015. *Pendekatan Pengajaran Matematik yang Dilaksanakan dalam Bilik Darjah.* Retrieved 21 May 2019, from https://www.slideshare.net/ChanChewYen/pendekatan-pengajaran-matematik-yang-dilaksanakan-dalam-bilik-darjah
[16] Buteau, C., Muller, E., Marshall, N., Sacristán, A. I., and Mgombelo, J. 2016. *Undergraduate Mathematics Students Appropriating Programming as a Tool for Modelling, Simulation, and Visualization: A Case Study. Digital Experiences in Mathematics Education*, **2(2)** , p 142-166.

AUTHOR INDEX